I I S S

STRATEGIC SURVEY

1995-1996

Published by  Oxford University Press for

The International Institute for Strategic Studies
23 Tavistock Street
London WC2E 7NQ

STRATEGIC SURVEY 1995–1996

Published by Oxford University Press for
The International Institute for Strategic Studies
23 Tavistock Street, London WC2E 7NQ

DIRECTOR
Dr John Chipman

EDITOR
Sidney Bearman

This publication has been prepared by the Director of the Institute and his Staff, who accept full responsibility for its contents, which describe and analyse events up to 29 March 1996. These do not, and indeed cannot, represent a consensus of views among the world-wide membership of the Institute as a whole.

Managing Editor: Rachel Neaman
Production Supervisor: Denise Fouché
Research Assistant: Bartholomew Goldyn

First published April 1996

ISBN 0 19 828 0912
ISSN 0459 7230

Strategic Survey (ISSN 0459 7230) is published annually by Oxford University Press.

The 1996 annual subscription rate is: UK £20.00; overseas $US32.00.

Payment is required with all orders and subscriptions. Prices include air-speeded delivery to Australia, Canada, India, Japan, New Zealand and the USA. Delivery elsewhere is by surface mail. Air-mail rates are available on request. Payment may be made by cheque or Eurocheque (payable to Oxford University Press), National Girobank (account 500 1056), credit card (Access, Mastercard, Visa, American Express, Diner's Club), direct debit (please send for details) or UNESCO coupons. Bankers: Barclays Bank plc. PO Box 333, Oxford, UK, code 20-65-18, account 00715654.

Claims for non-receipt must be made within four months of dispatch/order (whichever is later).

Please send subscription orders to the Journals Subscription Department, Oxford University Press, Walton Street, Oxford, OX2 6DP, UK. Tel: +44 (0) 1865 267907. Fax: +44 (0) 1865 267485. e-mail: jnlorders@oup.co.uk

Strategic Survey is distributed by Mercury Airfreight International Ltd, 10 Camptown Road, Irvington, NJ 07111-1105, USA. Second class postage paid at Newark, New Jersey, USA and additional entry points.

US POSTMASTER: Send address corrections to *Strategic Survey*, c/o Mercury Airfreight International Ltd, Cargo Atlantic, 10 Camptown Road, Irvington, NJ 07111-1105, USA.

PRINTED IN THE UK by Bell & Bain Ltd, Glasgow.

CONTENTS

Perspectives

Two themes stood out sharply in 1995–96 as states struggled to find the proper strategy for resolving continuing conflicts. Most important was the tribute paid to the nineteenth-century concept of balance of power. The US again stepped forward to lead. After crafting a peace in Bosnia, it helped to organise, under the NATO umbrella, a collection of states to divide and control the small but brutal Balkan forces that had caused so much misery for four years. When China tried to intimidate a smaller Taiwan, it was openly deterred from any thought of using direct force by a large US naval contingent rapidly and effectively deployed to the region. Those capable of forming 'national interests' can still have them shaped by an appreciation of the balance of power: in 1995, 'compellance' and 'deterrence' re-entered the strategic lexicon.

Added to this appreciation of the value of old techniques, however, was a recognition that the greatest world powers, and the effectiveness of the most astute diplomats, remain vulnerable to the vagaries and localised brutality of terrorist acts. The US and Japan both suffered irrational terrorist actions on their own soil. The peace processes in the Middle East, Ireland, and to some degree Chechnya, were set back by small groups of bombers and fighters who saw no need for diplomacy.

While relearning how effective the nineteenth-century game of balance of power is for handling traditional crises, states throughout the world will need to combine it with a twenty-first-century technique – yet to be elaborated – to deal with sub-state violence. Unless they can do so, much inter-state diplomacy will be destroyed by those who can still frustrate the peace plans of even the most devoted and powerful champions of diplomacy. Strategists need to balance the power of other states, but they must also deal with the power of those who act outside the state system. All too many of them see no value in the type of peace carved out for them by political moderates.

It was not only the yearning for peace that was unrequited in 1995. Hopes that countries could begin to cooperate more closely also suffered. The strength of support that the Russian Communist Party received in the December 1995 parliamentary elections could not be dismissed lightly. Although no longer driven by a fully communist ideology, the Party caused increasing concern through its expression, and exploitation, of extreme nationalist sentiments. The leaders of the Chinese Communist Party, perhaps partially because of the insecurity that an ongoing succession struggle brings, also worried their neighbours and the US by their aggressive hubris.

Fissures appeared in even the most stable societies. Japan's self-confidence was rocked, not only by a massive earthquake which destroyed what were thought to be earthquake-proof structures, but by an apocalyptic religious cult which released toxic gas into the subway system. The United States was awakened to the dangers of indigenous extremists by the massive explosion that gutted the Oklahoma City Federal Building, killing over 160. The poison of terrorist attacks had spread across the globe.

Perhaps belatedly, in 1995 the US reasserted its character as a super-power. It demonstrated that the discriminating use of force, or its threat, was a useful conjunct to diplomacy. Both making and keeping peace in Bosnia depended heavily on the judicious application of force. US support for the heavy bombing of Bosnian Serb positions, and for a Croatian offensive which virtually eliminated the Serbian military presence in Croatia and reduced Serb control in Bosnia considerably, brought the Serbs to the peace table. US President Bill Clinton then overcame Congressional and public reluctance and sent US troops under a NATO mantle into a possible combat zone. The Implementation Force (IFOR) was manned by heavily armed and highly mobile troops who threatened to respond with considerable force to any breach of the peace. This was enough, for the moment, to maintain the momentum.

In East Asia, the US steadied nerves in March 1996 when it sent two aircraft carrier battle groups close to Taiwan to 'monitor' the threatening Chinese manoeuvres in the Taiwan Strait. This show of force was intended to send the message to Beijing that the US will stand behind its commitment to support a peaceful, mutually agreed unification of China and Taiwan, but will not allow unification by force. It was a message China may not have liked, but could not ignore. Beijing continued to bluster about the inadmissibility of Taipei's purported independence moves and foreign intervention, but hastened to assure the world that it had no intention at this time to take the island by force.

The True Believers

Most countries, even those that appear as uncompromising as China, are headed by leaders whose bellicosity is tempered by a strong sense of self-preservation. They do not deliberately court the possible destruction of their country, or their own ability to rule, by pursuing a temporarily unattainable goal. They generally accept that it would be better to live to fight another day, using the interim period to improve the means and circumstances to produce the desired result. This, unfortunately, is not true of the spirit of the true believer.

Imbued with religious, nationalist or ideological fervour, his mind is closed to rational argument or thoughts of tolerance. His identity is subsumed into a greater whole; thus appeals to self-interest, or efforts to awaken him to the dangers of self-destruction, fall on deaf ears. In some cases, self-destruction is positively welcomed in the belief that it will result in personal aggrandisement in a better world far removed from this one. The true believer is unmoved by the fact that his actions kill or maim the innocent, man or woman, young or old. For him the 'enemy', either the state which refuses to redress what he considers wrongs, or some inchoate entity, is responsible. He is absolved by his righteousness.

In an earlier period, the true believer, like Savanarola, might have turned to preaching. Today, his chosen path is that of the terrorist. Terrorism is a weapon of the weak against the strong. The weaker and more marginalised the true believer is, the more likely he is to turn to a dramatic violent act to make his presence, and that of his cause, known. Reliance on such members leads to one of the dangers terrorist organisations face. If their leaders decide that new, more moderate tactics will forward their goal, while further destructive actions will only cause it to recede, the ensuing frustration of some of their members breeds frenzy.

These organisations then split, and sometimes split again, leaving the most extreme fanatics totally uncontrolled and determined to continue the good fight. It does not take many to cause havoc. This seems to have happened to both *Hamas* and the Irish Republican Army (IRA)/Sinn Fein organisations. Some of the leaders of *Hamas*, which has in the past relied on both social engineering among the Palestinians and terrorism in Israel to advance its cause, seem to have decided to explore the political option. The more moderate and pragmatic segment of the *Hamas* leadership even agreed to join in the elections that the Palestinian Authority was finally able to hold in January 1996, although earlier it had said it would never do so.

A more radical wing, proclaiming revenge for the killing of the notorious *Hamas* master bombmaker, Yehya Ayyash, by the Israelis, carried out a series of suicide bombings in early 1996 that took the lives of almost 60 Israeli citizens. It took some days for *Hamas* to accept responsibility for this wing. The moderates in *Hamas* may have felt that renewed violence would be a mistake. The need for unity, however, prevented a renunciation of the bombings. Once again, a more militant tail was wagging the more pragmatic dog.

The shock of these attacks threw the outcome of the national elections, called by Prime Minister Shimon Peres, into doubt. The Labour Party had lost its strongest asset, then-Prime Minister Yitzhak Rabin, to the bullet of a fanatical Jewish student. Now its political position and the peace that it championed were compromised. Most Israelis are prepared to trade land for peace, but only if that peace gives them personal security. They are less interested in Peres' grander vision of a comprehensive peace bringing diplomatic recognition, more trade and even increased tourist exchanges. If having gained all that they are still faced with the threat that they or their children may be blown apart by a suicide bomber, they will not accept the trade.

The peace in Northern Ireland also seemed to have been affected by a similar dynamic of frustration and violence. The leaders of Sinn Fein had apparently decided that continuing indiscriminate attacks would never move any UK government to accede to their demands, nor would violence help raise the political consciousness of the people of Northern Ireland. If anything, it was having the opposite effect. Either a part of, or the whole of, the militant wing of the IRA decided after 17 months of peace that they could no longer wait for the promised all-party talks that would give them at least a voice in determining policy for Northern Ireland. They also feared that they would be required to disarm, a prospect they refused to consider. None of the leaders of Sinn Fein, including Gerry Adams whose credibility was badly tarnished by the renewed violence, sounded very confident of being able to rein in the fanatic true believers.

Baiting the Bear

Although the single-minded nationalist rebels in Chechnya are unlikely to attain outright independence as they demand, they may well have a more significant effect on the wider Russian scene. As President Boris Yeltsin himself has said, unless he can find a way to put the Chechen affair behind him before the presidential elections in June 1996 he will have little chance of staying in office. Nevertheless, although they cannot expect any concessions from a communist government led by the extreme nationalist Gennady Zyuganov, the Chechens do not seem prepared to grant Yeltsin's wishes. He cannot afford to give them full independence; they refuse to give him peace.

The Chechens are masters of the unexpected. Fully prepared to sacrifice themselves for their cause, they are effective in hit-and-run battles against Russian conscripts who have no cause to believe in and thus no willingness to sacrifice themselves. For an overwhelmingly outnumbered, theoretically defeated rebel band to have attacked, seized and held for four days in March 1996 one-third of Grozny, the capital city that had been run by the Russian Army for almost a year, was devastating for Yeltsin. Another event like this could destroy his hope of remaining Russia's President.

If Yeltsin has any hope, it is that the other candidates appear worse than he. Most elections around the world are decided by negative votes.

The electorate scans the lists for the least bad choice, votes against the others and thus anoints a winner. Boris Yeltsin is hardly anyone's first choice. His health is not good. Two heart attacks in 1995, even if slight, raise questions about his ability to finish another term. His tendency to overindulge in drink, in histrionics and in periods of inactivity provide an easy target for the opposition. Yet his sterling credentials in helping to end communist rule, in moving towards a free economy, and in supporting democratic norms are hard to match. Even for an electorate that is frustrated by the slow effect of economic improvement on its own lives and is irritated by the loss of imperial greatness that it enjoyed not so long ago, Yeltsin may still be someone to gamble on.

The rest of the world has done what it can to help this least bad choice. Leaders of the West have given him rhetorical support, the World Bank and the International Monetary Fund have provided last-minute monies to fill his empty coffers, and the international community has been careful not to embarrass Russia to avoid making Yeltsin an easier target for nationalist attacks. These moves were all admirable and had to be taken, but they cannot be expected to swing the election. Assuming that the election goes ahead as planned, the fate of Russia will be determined by its own people.

The Russians have been remarkably patient and forbearing. That patience may now be wearing thin, however. The economic reforms that Yeltsin has backed, including privatisation and developing a market economy more open to foreign investment and joint ventures, has brought wealth to very few, but economic misery to many. Crime is out of control. Corruption is rampant. The long-suffering old with inadequate pensions, and the poor in the city centres and the benighted countryside, have been hardest hit. To them, acquiring the ability to speak their minds freely in exchange for order in the street, even if repressive, and for stability of income, even if on a low level, is a luxury they do not feel they can afford.

Many Russians have grown fond of democracy, however. In the parliamentary elections in December 1995 they used this tool to express their unhappiness with the administration by giving the Communist Party a majority of their support. But whether they are prepared to turn both the administration and the legislature over to the communists is in doubt. As the presidential election campaign unfolded, more questions were being asked about the direction the Communist Party intended to take the country. Few relished the thought that this first election for President might be the last.

Boris Yeltsin is a wily politician. Like Bill Clinton, he runs a far better campaign for office than he performs once in office. He demonstrated this again in February and March 1996 by abjuring responsibility for his country's economic troubles. Instead, he blamed the incompetence of his reform-minded colleagues, and then sacked them. With the stroke of a pen, he fulfilled the peasants' centuries-old desire to own the land they till. Without trying to match the extreme nationalism of his opponents, he has put himself in a position to benefit from the re-awakened fervour for Russian greatness. It may not be enough, however, to overcome the opposition's trumpeting for a return to the USSR, enveloping the independent nations that have emerged from the destruction of the Soviet Union in 1991. Hopes that an easier relationship would develop between Russia and the West may be shattered by the revitalisation of nationalist sentiment and longing for the past that has swept over Russia.

Agitating Eastern Waters

China, too, is busy disappointing those who thought it might have become more amenable to accepted norms of international behaviour. Since Deng Xiaoping launched his economic reforms, China has agreed to international investment and aid to provide the impetus for a modern industrialised state. It was hoped that this growing economic interdependence would temper China's more aggressive tendencies. Instead, as China grows stronger, it grows more bellicose. Its insistence on having things its own way has begun to deeply concern even its erstwhile friends in Asia.

While allowing its people greater opportunity for economic participation in order to unleash their genius for capital accumulation, China's leaders are adamant about maintaining a political dictatorship. They are determined to keep themselves in power, and power, Mao famously said, grows out of the barrel of a gun. For contemporary leaders in China it will certainly never grow out of a ballot box. The dilemma that is posed, then, is how to motivate and keep together their sprawling, dynamic country. Communism, as an ideology with a unifying theme, died in the rush to develop market capitalism. Democracy, in the sense of allowing the people true participation in political affairs, is anathema. What is left as a useful unifying force is nationalism.

China's nationalism is particularly virulent. It is profoundly anti-Western, fuelled by atavistic memories of a time when China was the centre of its world, recognised and treated as such by weaker neighbours. This idyll was shattered by the appearance of militarily vastly stronger civilisations from far away, which pushed the weakened dynastic rulers around, infected the country with pernicious foreign ideas and even seized chunks of territory. The Chinese have neither forgotten nor forgiven. As they grow stronger they intend to right what they see as wrongs, redress the slights – many of them real, some imagined.

With this background it is easy to understand and explain Chinese bellicosity over Tibet, Hong Kong and Taiwan. To Chinese leaders, and their people, these are areas torn by force from the homeland by meddling foreigners. The Chinese intend to fold them back into that homeland, by peaceful means if possible, but by force if necessary. Tibet has already been reabsorbed, and Chinese sovereignty is no longer really questioned by any but the Tibetans themselves. Hong Kong will return to China's control in July 1997. Taiwan, however, is a tougher morsel to swallow.

China's behaviour can be understood, it cannot be condoned. Threats to use force against the Taiwanese, who have no desire to be reunified with China, must be opposed. Taiwan has developed a vibrant free economy and newly thriving democracy which have given its people greater self-confidence. When that self-confidence is expressed in too strong a desire for independence, however, Taiwan's friends must be careful not to appear to either Beijing or Taipei as encouraging it. The Taiwanese deserve more involvement in international affairs, but if they push this too far they will drive the Chinese to hostile acts that will inevitably involve other powers. The line that must be drawn is a delicate one and requires adroit diplomacy.

In the crisis created in March 1996, when the Chinese mounted massive manoeuvres in the Taiwan Strait to intimidate the Taiwanese, the US reacted adroitly. It moved in sufficient force to convince China that it was prepared to take any necessary military action to block Chinese force, while at the same time reiterating its policy that the long-term goal must be a peaceful settlement between China and Taiwan. Only the US could have played this essential balancing role, and it was tacitly welcomed by China's Asian neighbours. They are very aware that their efforts to provide security in the area through a consensus which all the nations, including China, will embrace and live up to requires a long incubation period. In the interim, a consistent US policy to support their desire for peace and stability is very welcome.

The most recent Chinese effort to intimidate Taiwan seems to have failed. In the 23 March Taiwanese elections, the people – 76% of those eligible – poured out to vote and gave President Lee Teng-hui a 54% majority, larger than originally forecast. Their second choice was the candidate whose party has been insisting on Taiwanese independence. China appeared to have grasped that for the moment its hostile attitude had brought unexpected results. If anything, it had solidified Taiwan's move towards democracy. The crisis in the short term has abated, but the future is still uncertain. The question that remains, and may have grown more acute, is how the two countries will ever accommodate their contradictory goals.

Waiting for a 'generational change' in China to bring positive progressive changes in its outlook, as is often suggested, is hardly the answer. History is rife with examples of 'generational change' that brought regressive negativism in their wake. To be younger should not be equated with being better. What is required, no matter in what generation, is a change in thinking and the acceptance of a new reality. Deng Xiaoping, after all, was the same generation as Mao, and it was he who instituted the sweeping changes that have begun to revolutionise Chinese life. These economic reforms provide the best hope for the right kind of change.

China's leaders, including Deng, believe that constant economic growth will provide such well-being that the people will continue to support the regime. To a degree, this is correct. But only to a degree. For tension has developed in China between the need to allow freer thinking and freer action in the economic sphere, and the leadership's desire to prevent any expression of such freedom in the political sphere. China can no longer restrict knowledge to a favoured few, however. Many need to know much if the economy is to continue to thrive.

The examples of South Korea and Taiwan, where democracy is replacing autocratic regimes without slowing economic growth, have given the lie to the oft-expressed view that in Asia solid economic growth requires autocratic central rule. If there is to be a change in China, it will come as a result of weighting the scales in the direction of greater political participation to ensure continuous economic development. Only when China and Taiwan draw closer in political views, as they have in economic relations, can greater integration be expected. In the meantime, however, China is certain to continue to cause trouble in the seas around Taiwan. In response, the US needs to keep its ships as close as Japan, and its powder dry.

The Only, if Reluctant, Superpower

The US show of force in Asia was greeted in the region with public silence, but private satisfaction. In Europe, US use of force in Bosnia received the opposite response – public relief, but private chagrin. Although US military and diplomatic pressures had led to a cease-fire and a peace agreement among the three warring Bosnian forces, its allies in Europe felt that the US had joined in too late, had usurped European ideas and claimed them for its own, and had been overbearing in its insistence that everything had to be done its way or it would pack up and go home. The Europeans had a good point – up to a point. But their ideas had been ineffective because of lack of coordinated action for almost four years. It required US single-minded direction and military force to silence the guns in Bosnia, and if the price to be paid was the loss of some European face, it was a price worth paying.

President Bill Clinton's new-found activism in foreign affairs was not as great a departure for him as is sometimes asserted. He had not replaced his interest in domestic affairs with a new foreign-affairs agenda. Instead, much of that agenda was driven by domestic political considerations. The Republican Party majority in the Congress had reduced his ability to take positive action on the home front, leaving him with few weapons but the power to veto legislation that he did not like. The leaders of Congress were also reaching beyond their interest in bringing major changes on the domestic scene to play an unhealthy role in foreign affairs.

Thus the foreign crises Clinton chose to deal with were also essentially blocking actions within a domestic political framework. The Republicans in Congress threatened independent action over China that would have had deleterious effects on US Asian policies for years to come. Their views on Bosnia threatened to disrupt the fundamental transatlantic NATO structure. Even Clinton's activist policies with regard to Northern Ireland came in reaction to Congressional pressure, even if in this case the pressure was generated by Democratic Party leaders of Irish extraction. If in 1995 the US was again acting as a superpower, involved in putting out fires around the globe, it was not because it had created a new, consistent foreign policy.

Even if most of it was Clinton reacting, he was clearly pleased by the accolades that accompanied his apparent successes. Putting out fires, however, is a long-term procedure. The fire fighter must remain long enough to ensure that the embers have stopped smouldering or the fire rekindles. This has already proved the case in Northern Ireland and the Middle East, and it threatens to happen in Bosnia, East Asia and perhaps even Haiti as well. Even if the US is a reluctant superpower, to make a useful difference it must remain a persistent superpower. It would be disastrous for the world if the US were to retreat to the kind of isolation some Republican aspirants for the presidential nomination espoused.

What Role For Europe?

In the specific case of Bosnia, however, the fact that Clinton faces an election campaign and that the Republican Party has made the date for US forces to withdraw a *sine qua non* for allowing their deployment in the first place means that withdraw they will by the end of November 1996. If the European troops that help make up IFOR follow, the delicate peace structure is likely to collapse. This would be more than a shame; it would be a disgrace. Bosnia is still a European problem, as it was when the fighting began more than four years ago. Little of value was done about it then, and there is no excuse for allowing it to happen again.

The part that NATO finally played in convincing the three Bosnian parties to stop the killing was significant. In the contemporary world, violence is all too often an instrument of evil. The NATO action showed that force can still be used judiciously for good. Through its involvement, NATO has also rediscovered a sense of purpose. Its engagement in Bosnia has served to knit back together the fabric that loosened when Charles de Gaulle led the French out of NATO's integrated military command structure some 30 years ago. Current President Jacques Chirac has brought the French military back, partially because the role France has been playing in Bosnia now meshes well with NATO. It also offers an opportunity for NATO troops to work closely with Russian forces who, by clever sleight-of-hand, are attached to the IFOR organisation while not being part of it.

Much of this may be lost by the end of 1996 if agreement is not reached to renew some part of the peacekeeping/peace-enforcement mission. It would perforce be smaller. Yet it would have to be sufficient to hold the peace together until the new structures begin to gel. For their own sake, as well as for those who have been suffering in Bosnia, the European states must show that they can lead as well as follow. Much of the credit for the initial conditions for peace has gone to the Americans. Maintaining the peace over the longer term, and building a more solid structure upon it, can devolve to Europe's credit, if it can get its act together. It is past time for it to do so.

Other areas in which Europe had hoped to cooperate have begun to look less encouraging. Efforts to bring about Economic and Monetary Union (EMU) stumbled badly against the inability of most countries to meet the criteria laid down by the November 1993 Maastricht Treaty. In both France and Germany, the two countries apparently most in favour of an EMU, the citizenry, and some leaders, are rethinking the proposition. This is clearly a path that Europe should not tread unless its people have expressed their real support. Pressure has been growing for referenda to test that support. This is no doubt a sensible precaution, even if the result shows that not all European countries are prepared to join. One possible outcome is that the date for EMU will be moved further into the future, while adjustments to the way in which an EMU can be effected are made.

Nor is this the only likely postponement. Preparations for the March 1996 Inter-Governmental Conference (IGC), intended by some to strengthen the EU's ability to act more as a whole, did not augur well in 1995. No agreement was made on such basic issues as extending qualified majority voting, on which nations should have a veto on crucial foreign and security decisions, on whether more power should be given to the European Parliament, or on how to adjust and revise the Maastricht Treaty. The UK, as always on matters European, stands out in its opposition. But it is not alone on all issues.

Europe's main difficulty lies in deciding whether it should move first to tighten the European arrangements, as the Germans want, or enlarge the membership, as the British want. This basic difference of opinion threatens to block all forward movement. It will probably result in the Conference lasting well beyond the 12–15 months originally envisaged. Some European states may hope that before the end of the IGC, elections in the UK will bring a Labour Party to power that will be more amenable to the views of Europe's majority. This is doubtful. The Labour Party may support some proposals, such as the Social Chapter, but is no more likely to risk getting ahead of public opinion, which does not support fuller integration, than the Conservative government. Those intent on driving forward may have to be content with a Europe built on a variable, multi-speed model. This is not quite the vision that has been held out for the last ten years, but it is another small step forward.

One Forward, Two Back

Progress in human affairs has always been uneven. At times it speeds up and for every step backwards, there are two steps forward. At other moments it appears the reverse. The 15 months since January 1995 would seem to have been part of the latter cycle.

The United States had been badly missed as the vital power for world peace and stability. In 1995 it returned to the fore in world affairs. Its resurrection of a balance-of-power role can only be welcomed, but sustaining that position will not be easy. The American people are in two minds about extending their power beyond their own shores. If vital interests are at stake, support will be easily found. It can easily be lost, however, if the many loud voices proclaiming narrow self-interests convince the people that the only vital interest the US has is the security of its own borders. For this, the President must continue to lead as he has during the year. In the 1996 election year, this cannot be assured.

The US has demonstrated that balance-of-power policies can work to redress imbalances when the opposition is clear and well defined. In the Balkans and in Asia, US ability to project power, and a demonstrated will to do so, was successful. In meeting what has become an equally serious terrorist challenge to peace, however, this is not sufficient. To protect a state's citizens from random bombings is difficult enough; to prevent those prepared to blow themselves up in pursuit of their cause is doubly difficult. Yet, if some answer is not found, the world will find itself hostage to its least civilised, least tolerant elements.

But developing an alliance against terrorism will be exceedingly difficult. A start was made in the Middle East with the 'meeting of the peacemakers' that assembled in Egypt in mid-March 1996. Twelve Arab leaders joined with the United States and Israel to condemn the recent suicide attacks by Islamic militants and to pledge an exchange of information to try to root out the terrorists before they strike again. This expression of cooperation in a part of the world that has long been known more for its conflict can only be encouraging. The need to go beyond broad pledges to joint concrete actions, however, is still very much required.

Nations can do much to integrate their efforts against terrorists, but it will require an effort by the mass of generally silent people to really make the change. Their desire for a peaceful and secure life must overcome the instinct to turn a blind eye to the terrorist in their midst. If the safe haven they supply is removed, the terrorists will have such difficulty hiding that the state will be able to use its available means more effectively and turn the tide. This may be happening in Northern Ireland, where the people on both sides of the divide have indicated their dismay at the return to violence after 17 months of peace. Without this fundamental change, the nations of the world may find that regulating affairs among themselves, even with nations as prickly as China that insist that only they have right on their side, is a less daunting task than preventing the weak and disaffected from destroying hopes for peace and security.

Strategic Policy Issues

Defence Industries In Transition

The international defence industry is in a state of unprecedented turmoil. It is increasingly apparent that previous cutbacks and mergers, dramatic though they seemed when they occurred, have only postponed a much more fundamental restructuring. In many countries, major defence producers, once the politically untouchable stars of national industrial development policy and symbols of national autonomy, are leaving the defence business or closing down altogether. Others are shifting their orientation to purely civilian markets, leaving only a handful of firms to dominate the increasingly speculative defence sector. Is the defence industry undergoing consolidation, or is the process closer to complete transformation tinged with the real danger of collapse?

As threats of inter-state war recede, major defence industries, previously perceived as an essential part of modern states, have lost much of their importance. Many governments find that costs they readily bore for decades are increasingly unacceptable. Highly publicised mergers in the US, the decline of major manufacturers in much of Europe, and the near collapse of production in Russia and much of the Third World paint a stark picture indeed. Except for a handful of US giants, the future of the defence industry as a whole is increasingly bleak. While some producers will find security in niche markets and others will profit through foreign sales, most will require substantial reorientation if their vital role in national security is to continue.

So far, the only certainty is that the defence industry has lost its unique and special nature, and is yielding more and more to the ordinary rules of the civilian market-place. As defence spending declines, no government can afford to subsidise its defence manufacturers with large long-term orders at inflated prices. Instead, defence businesses, like their civilian counterparts, are now run more on the basis of efficiency and competitiveness, emphasising short-term flexibility and responsiveness to business opportunities. For defence firms of the future, long-term contracts will be few and far between; the ability to react swiftly and improvise may now often be more important to a firm's well-being.

But military suppliers will never be able to play by exactly the same rules as those in civilian markets, if only because military equipment is purchased and used so differently. Even in their reduced form, contemporary armed forces will continue to require new equipment, some of it unrelated to anything available on the civilian market. With the industry in turmoil, it is not clear where this equipment will come from. The great dilemma for military-industrial policy will be how to preserve the unique capabilities of advanced defence industries without paying the kinds of sums invested in the past. How can a nation develop and produce state-of-the-art military equipment even when it is not buying any? Are there alternatives besides relying on the United States or paying enormous sums to subsidise domestic capabilities?

The Changing Industrial Environment

The difficulties faced by international defence industries were greatly exacerbated by the end of the Cold War, but the roots of their problems go back much further. As early as the 1950s, the rising costs of new technologies – such as the turbojet engine, missiles, electronics and advanced materials – forced even the best-funded armed forces to choose between quality and quantity. Gradually, most countries allowed their armed forces to shrink to pay for modernisation and to respond to a new security environment. By the 1970s, responses became increasingly uniform. In country after country armoured divisions have been replaced by brigades, air wings by squadrons, navy cruisers by destroyers and destroyers by fast-attack boats. Through the initial years of this process, the pretence could be maintained that any sacrifices were illusory; size and brute destructiveness were being traded for greater capability, precision and effectiveness.

In the 1990s, however, it has been impossible to hide the erosion of actual military capability. Even before the end of the Cold War, all but a handful of European countries and major actors elsewhere permitted their armed forces to decline. In Europe, the cuts in conventional forces mandated by the 1992 Conventional Forces in Europe (CFE) Treaty have broadened and institutionalised a process already under way. In other regions, like Latin America and Africa, defence spending has been falling even more rapidly, as has the number of men and women in uniform. Many military units have been disbanded because they are no longer needed and to economise, while modernisation has been delayed, stretched out, or limited to a handful of elite units, and costly training has been minimised.

The assured profits of defence production have also been lost. Those defence firms with civilian sectors typically find that these sectors now support their military divisions. In the highly visible aircraft sector, for example, aerospace analysts now expect the world's air forces to buy approximately 3,500 new fighters over the next 20 years, while civil airlines will buy some 15,500 commercial airliners.

As the demand for military equipment declines, reduced buying has a direct effect on suppliers. Although most industrial nations have tried hard to keep military procurement budgets as high as possible and to protect defence-industrial capability, their ability to shelter this favoured sector is increasingly limited. As global defence spending began its sharp decline in the late 1980s, falling from $1.2 trillion in 1987 to $850 billion in 1994 (in 1994 dollars), the effects of market decline could not be avoided. The chaos of the changes in the former Soviet Union brought much of its defence industry to a virtual halt. Firms in Europe, North America and most of the Third World reacted in much the same way as contracting manufacturers in other fields; they postponed capital improvements, trimmed work forces and questionable research and development (R&D) programmes, aggressively searched for export opportunities and diversified into new product areas. Only the most protected industries in a handful of states like China, India and North Korea have been immune to these pressures.

This process of slimming and rationalisation reduced defence industry employment dramatically, from 1.6 million in Western Europe alone in 1990 to 900,000 in 1995. Despite the considerable pain involved, combined with concomitant gains in efficiency, there were clear limits to this process. Although the name and ownership of some defence firms changed, no major Western manufacturers left the defence business. Even more revealing, no large production programmes ended prematurely (although several development projects did, such as the UK's *Nimrod* maritime reconnaissance aircraft, the NATO Frigate Replacement and the US Navy's A-12 attack aircraft).

In the United States, these sales and mergers allowed firms to reorient their business, as when General Dynamics sold its tactical aircraft and tank business and General Motors sold *Hughes* missiles. In Europe, a few new multinational firms appeared, most importantly Eurocopter SA created by France and Germany in 1991, but these still have distinct national divisions, and mergers were more commonplace. In Germany and France, mergers were limited to particular sectors: Deutsche Aerospace took control of practically its entire sector, the French land-forces manufacturer GIAT absorbed all similar French firms, and the French electronics firm Thompson-CSF bought related firms in France, as well as in the UK and the Netherlands. In the UK, GEC and British Aerospace assumed commanding shares of the nation's defence sector, and Italy's state-controlled Finmeccanica acquired an even more dominant position by purchasing several other state-owned enterprises.

For all this activity, the basic rules of the defence business have remained essentially unchanged since the 1960s when the last major innovations came through establishing multinational programmes like *nado* and modern co-production with offsets. These mergers did not generally lead to closing redundant subsidiaries or integrating disparate operating divisions. On the contrary, their principal effect – and often their deliberate goal – has been to preserve weak enterprises by facilitating temporary cash injections. One sign of the modesty of these seemingly profound changes is the continued survival of highly criticised projects such as the Anglo-Italian EH101 helicopter, Italy's *Ariete* main battle tank, the UK's *Tornado* Air Defence Variant or the RDM radar for France's *Mirage*-2000 fighter.

How Have Nations Responded?

The industrial restructuring that began in 1994 has been of a different order. Restructuring was perhaps most profound in the United States, although its implications have been felt globally.

The United States

The international defence industry will be revolutionised by a new wave of industrial concentration symbolised by the merger in August 1994 of the aircraft-maker, Lockheed, and the missile and electronics firm, Martin Marietta. The dominance of the merged firm was further assured in January 1996 when Lockheed-Martin bought the defence electronic and guidance divisions of Loral for $9bn. In the United States, previous defence conglomerates had been created by acquiring essentially unrelated divisions – like General Dynamics, the largest US defence firm of the 1970s and 1980s in tactical aircraft, shipbuilding and tanks. The creation of Lockheed-Martin, with defence sales of approximately $20bn annually, was widely acknowledged as the harbinger of a new phenomenon, the vertically integrated defence giant with near-domination of broad military-industrial sectors.

The sheer size of the US defence market – still double the value of all European defence procurement – combined with the market coherence of a single major buyer gives large US firms enormous advantages over their rivals. Unlike state-owned or legally restrained firms in Europe, US firms are free to cast out redundant workers and divisions, while shifting the survivors to ensure industrial efficiency. With production facilities around the country, the new firms have substantial political influence. Because they deal primarily with a single government, they can sign contracts much faster than their European counterparts, while being certain of much larger and more economic production runs. Through sheer size they are unsurpassed in private R&D resources. This new defence enterprise has the further advantage of fully integrated R&D work, pooling the activity of all its divisions to maximum effect, minimising overlap and other inefficiency.

Similar mergers have brought together the remaining resources of Northrop-Grumman, which purchased the defence electronics business Westinghouse in January 1996 for $4.2bn. Boeing and McDonnell-Douglas have also negotiated over a possible merger. These mergers not only strengthen the concentration of the US aerospace industry, but also facilitate cooperation between makers of platforms (air-frames) and sub-systems (like navigation, fire control and armaments). Their strength in electronics provides new capabilities to identify and take advantage of the military applications of civilian electronics technologies.

These mergers have not been without risk. To finance them, firms have allowed their average debt-to-equity ratios to rise from 36% in 1993 to 112% in 1996. For Lockheed-Martin, the ratio has risen to an estimated 179%, while for Northrop-Grumman the figure is approximately 265%. These risks are based on the assumption that the relatively favourable atmosphere for long-term defence contracts will continue for at least two decades. But the future of programmes such as the C-17 transport aircraft, the F-22 stealth interceptor, the Joint Advanced Strike Technology (JAST) attack fighter or the Theater High Altitude Area Defense (THAAD) ballistic-missile defence system cannot be taken for granted. The new conglomerates are highly vulnerable to the loss of particular programmes like these, programmes which may not survive new pressures to balance the US federal budget after the November 1996 presidential elections.

Western Europe

In Europe there is universal recognition of the significance of US and other challenges, but there is still no consensus on how to address them. Problems on the political agenda since the 1960s have yet to be resolved. Above all, Europe has a surfeit of manufacturers, firms that are promoted as national champions but that compete counter-productively in an era of accelerating integration. Although its defence market is half the size of the US, the European Union has ten aircraft makers against five in the US, 11 missile manufacturers against five, ten armoured vehicle makers as opposed to two, and 14 warship builders against four. When the Soviet-style arms industries of Eastern and Central Europe are included, the inefficiencies are even more striking.

The fragmentation of Europe's defence industry will not be on the agenda at the EU's March 1996 Inter-Governmental Conference (IGC), which has decided for the time being not to discuss collaboration in armament manufacture. Article 223 of the 1957 Treaty of Rome, which exempts defence contracting from the single market, is too sensitive for revision. But the inefficiency of the European defence industry has made it impossible to avoid the issue altogether. Concerns about the future of European defence collaboration were raised by the failure of the Western European Union (WEU) in October 1995 to agree on how to establish a European armament agency as planned. Instead, German Chancellor Helmut Kohl and French President Jacques Chirac agreed independently on 7 December 1995 to establish a new joint organisation, the European Armaments Agency, to manage all their military procurement programmes along with those of the Netherlands and Italy.

So far the main effect of the new arrangement has been to highlight French unhappiness with UK procurement policy. Until December 1995, UK membership had been blocked by France, where there is considerable resentment of the UK's preference to procure weapons on the basis of open competition rather than using the European industry. This long-standing dispute was highlighted in July 1995 when the British Army turned to the United States for *Apache* attack helicopters worth $3.7bn, rather than the Franco-German *Tiger*, provoking strong protests from France. A few days after the Kohl–Chirac summit, France permitted British Aerospace and Dassault Aviation to establish a joint venture to investigate a future joint combat aircraft to replace the *Eurofighter* and the French *Rafale*. But substantive cooperation will have to await the resolution of more basic disputes. The UK's willingness to join the next major new collaborative project, the $4.6bn Multi-Role Armoured Vehicle (MRAV), may determine France's position.

The established response to duplication is collaborative weapons procurement, pioneered by projects like the *AlphaJet, Jaguar* and *Tornado* aircraft. But the economics of such undertakings have always been controversial; they ensure larger production runs, but with greater unit costs and lengthy delays. Collaborative programmes have been unpopular with military commanders, who felt they were paying too much for weapons that did not serve their needs. Their greatest contribution has always been to political cooperation among participating governments. The problems became especially vivid in the current $60bn *Eurofighter* programme as Germany, led by Defence Minister Volker Rühe, tried to withdraw from the project. Forced to stay in for the sake of political relations with the UK, Italy and Spain, Germany cut its planned purchase from 250 to 140 aircraft, but sought to keep its workshare as originally negotiated.

The political role of weapons collaboration makes these the safest defence projects in Europe today. But there is widespread recognition that future collaboration must be much deeper, assuring not just shared participation in common projects, but also accelerating progress towards a fully integrated European defence industry. France and Germany are spearheading efforts to collaborate to minimise duplication and surplus capacity. Under a complicated trade-off agreement reached in December 1995, for example, Germany has become a major partner in France's $2bn plans for the *Hélios*-2 and future *Horus* spy satellite programmes. In exchange for DASA's leadership of a new enterprise, European Satellite Industries, France's Aerospatiale will assume the leading role in a parallel missile enterprise, European Missile Systems.

The most dramatic cuts in industrial capability occurred almost without discussion in the smaller European countries. Countries like Belgium, the Czech Republic, Slovakia, the Netherlands, Poland, Romania, Sweden and Switzerland no longer try to maintain the comprehensive capability to develop, or at least manufacture, all their military equipment. The same can be said of some larger European actors like Spain.

Since the 1960s, these countries have quietly allowed prominent segments of their defence industry to atrophy, moving from domestic design and development to co-production and, in the end, to importing finished equipment directly. Finland and Switzerland, for example, previously co-produced or assembled virtually all their tactical aircraft. But currently they are importing McDonnell-Douglas F/A-18 fighters in nearly finished condition, having abandoned their long-held ambition to maintain a domestic aircraft industry. Sweden continues to build its new indigenous fighter, the JAS *Grippen*, developed and built by SAAB Aircraft, but has abandoned its main battle tank capability in favour of importing second-hand tanks and armoured personnel carriers from Germany.

The collapse in March 1996 of the 77-year-old Fokker NV, the Dutch aircraft firm, was far more spectacular. Fokker's last exercise in military aircraft production ended with the co-production of F-16s with Lockheed-Martin in the mid-1980s. Since then the firm had eked out an increasingly unprofitable living by relying on its commercial aircraft and on maintenance. In 1993 Fokker was purchased by Germany's DASA. This was a typical European effort to save a small firm by giving it to a larger one, distinguished in this case by its multinational aspect. The arrangement's failure was acknowledged by DASA after the German firm endured steady loses as a result of the take-over, culminating in charges of DM2.3bn in 1996 for writing off its investments. That Fokker was allowed to collapse suggests that some European governments and industrialists believe they can no longer afford mergers if their primary purpose is to preserve weak enterprises. They may be more willing to permit consolidation of defence-related industries by allowing weaker branches to die naturally.

There is no evidence of equal *sang froid* among the major European powers. President Chirac's decision in February 1996 to cut France's armed forces came after three years of debate over the future of the nation's independent capabilities. But the implications for France's defence industries remain obscure. As the European power with the smallest stake in collaborative defence projects, French defence industries are exceptionally vulnerable. Their safety behind the nation's historic commitment to strong independent capabilities has been brought into question by cutbacks and by France's announcement in December 1995 that it would re-join NATO's integrated military command. The initial cutbacks were announced along with a decision to withdraw from the collaborative FLA transport aircraft and sharply reduce purchases of NH-90 and *Tiger* helicopters. Major defence producers were to be re-formed as well, including privatising Thompson-CSF and merging Dassault and Aerospatiale.

The industrial aspects of Chirac's announcement have met with opposition. German Defence Minister Rühe was especially forthright in his criticism of France's decision unilaterally to withdraw, or substantially reduce, its involvement in collaborative projects while its purely domestic programmes were left largely untouched. French industry has also shown that it may not cooperate. The Chairman of Dassault Aviation, Serge Dassault, has not hidden his distaste for merging his economically viable firm with the much larger and more troubled Aerospatiale.

In the short term, some firms have shown that much can be accomplished through sheer good management. British Aerospace was on the verge of collapse in 1991, with its civil and military business in sharp decline. The firm recovered to become perhaps the strongest major European defence enterprise in 1995, partially through exports – especially to Saudi Arabia – but also by aggressively laying off staff (by 90% in some divisions) and identifying numerous smaller business opportunities. Its experience shows that defence firms can prosper at least temporarily without major new contracts. In the long term, however, more ambitious measures will be essential to guarantee the long-term health of Europe's defence firms. European Union officials hope that the United States may still come to the rescue, arguing that its six-to-one surplus over Europe in defence trade is not politically sustainable. But even the largest foreseeable export orders cannot compensate for badly needed reorganisation.

Russia and Eastern Europe

The rapid decline of the ex-Soviet military-industrial system appears to have passed the point of technical collapse. With only a handful of exceptions, the broad capabilities that sustained the Soviet Union throughout the Cold War have virtually disappeared. Russian engineering teams retain the ability to promote highly advanced weapons concepts and, in some cases, they can delineate innovative designs, but the ability actually to produce sophisticated new weaponry has all but disappeared.

What remains is a huge industrial potential to mass produce weapons of 1970s and early-1980s vintage. Some of these systems, such as the MiG-

29M and Su-27/35 fighters, the S-300 air-defence system and numerous artillery systems, will be militarily potent for many years to come. Many allegedly new designs have been introduced since the collapse of the Soviet Union in 1991, but they are almost always modifications of existing weapon systems.

A typical example is the T-90 tank which was reportedly deployed in Russia in 1996. This is an updated version of T-72 tanks, the same tanks that were soundly defeated in battles between Israel and Syria in 1982. Even though the makers stress its new fire-control electronics, it is at best an interim system intended to justify continued production work at Russia's huge Nizhni Tagil plant. Development is proceeding on a completely new tank and on other systems, including a stealth fighter, but there is no evidence of the industrial wherewithal to take such plans beyond the prototype stage. Military systems that were still under development at the time of the Soviet collapse in 1991 – such as the A-50 *Mainstay* airborne warning and control system (AWACS) – have yet to be fully developed.

Of greater relevance is Russia's continued ability to mobilise existing productive capabilities. With its entire defence budget for 1996 planned at 80 trillion roubles ($16.7bn) there are minimal funds for production. Even after supplemental appropriations, only 20 trillion roubles ($4bn) are available for procurement. Much of this budget must be spread thinly to sustain over 5 million workers still employed by over 2,000 defence plants. Russian experts have concluded that sustaining even current rates of activity would require raising this figure to 50 trillion roubles ($10.5bn). In lieu of such financing, production declined by a total of 75% from 1991 to early 1996.

One of the most important sources of new technology and development funds is Western assistance, whether in the form of Nunn–Lugar-sponsored science and technology centres or collaborative projects for related high-tech fields like the NASA-led international space station agreed in 1994 and a 1996 test programme involving the Tu-144 SST. Another irony of this extremely depressed situation is the great sensitivity to even relatively modest export sales. The ongoing transfer of 18 MiG-29s worth perhaps $600m to Malaysia, for example, is of immense importance in sustaining the aerospace industry. Conversion to civilian production has been widely attempted as well, but is hindered by internal rigidity and by the general weakness of the Russian economy.

The situation is much the same in Ukraine, the only other state to emerge from the collapse of the Soviet Union with broad military-industrial capabilities. Here, too, industry remains in a state of suspended animation. Much attention focuses on the nation's industrial flagship, the Dnepropetrovsk Yuzhnoe and Yuhzmash ballistic missile and space-launch vehicle plants, the source of the SS-18 and SS-24 inter-continental ballistic missile (ICBM) as well as the *Zenit* and *Tsyklon* space-launch vehicles. Work on new ballistic-missile design reportedly continues. Efforts to broaden the civil market for Ukraine's *Zenit* space booster are slowly improving given its planned use to support Russian participation in the International Space Station and Ukraine's involvement in the Boeing-led Sea Launch Consortium for commercial space launchers.

Many Eastern European defence industries can trace their origins back to the inter-war years or even to the nineteenth century, but during the 1950s all were converted to Soviet-methods and designs. While the Czech Republic, Poland and Romania developed independent technical capabilities in niche areas, none had the ability to maintain their previous roles without Russian support. With their defence spending down to 20–25% of 1980s levels and procurement funds even smaller than before, their defence industries have all but collapsed. The exceptions are particular sectors like Polish helicopters and Czech jet trainers. Slovakia's armoured vehicle industry, centred at Martin, benefits from the uniquely strong political support of President Vladimir Merciar. These sectors receive sufficient support for incremental technical improvements to keep their products militarily useful. In a unified European arms market, however, they could only hope to survive through political largess.

China and the Third World

Where defence spending has not fallen and governments continue to support their defence industries directly, firms still play by the old rules of defence production. So long as money is plentiful and pressure for technical modernisation is minimal, they can continue to produce equipment designed in the 1950s and 1960s. Individually, such weapons are of dubious use on modern battlefields. But, as they accumulate, their sheer number may be sufficient to influence strategic perceptions. In specific situations such as remote or less intense conflicts, moreover, such veteran weapon designs may retain much of their combat effectiveness.

The nation that comes closest to this style of technologically detached military industrialisation is China. Even China has a strong interest in modernisation, however, and now faces growing pressure to cut its forces for this purpose. In some areas, such as its new solid-fuelled ballistic missiles like the short-range M-9/11 and the DF-31 ICBM, it has already modernised. In other fields, like tactical aircraft, China has purchased component technology and small numbers of finished systems, such as Russian Su-27 fighters and possibly *Kilo*-class submarines. Through such transfers China supports advanced domestic projects like the F-10 multi-role fighter, preserving the possibility of broad technical modernisation in the future.

The expense, however, may not be bearable. Analyses published in 1995 confirm the widely accepted belief that China's arsenal will shrink over the coming decade as weapons based on 1950s technology have to be retired. One-for-one replacement of its 4,000 fighters (3,000 of them copies of the MiG-19) and 8,000 tanks (6,000 based on the Soviet T-54) will not be possible. Like other developing countries, China is expected to emerge with a force structure divided between large units of marginal use and smaller numbers of more modern highly mobile formations. This leaves its industry with the choice of either accepting technical irrelevance, or introducing major cutbacks.

Most emerging regional powers have come to this conclusion already. Regional defence industries, which seemed to be growing throughout the Third World in the 1970s and 1980s, have been contracting and closing in the 1990s. Prominent projects, like Israel's *Lavi* fighter and Brazil's *Osorio* tank, were cancelled, while others, like Brazil's AMX and Taiwan's *Ching Kou* aircraft, will end soon after their truncated production runs.

Erstwhile regional military-industrial leaders like Argentina, Egypt and Indonesia have largely given up their defence production. Others, like Brazil, South Africa, South Korea and Taiwan, have allowed their defence industries to shrink around core areas. Some, like Pakistan, now mostly produce ammunition and other defence consumables. Only a handful still try to maintain broad military-industrial capabilities. The most prominent are India and isolated countries, for example, North Korea and possibly Iran. Like China, they are able to maintain broad military industries only by continuing to pursue aged technologies.

Selling Weapons Abroad

The obvious alternative for defence manufacturers confronted by the loss of traditional markets is to export. Three times in recent years observers have anticipated that the changing arms market would lead to an export frenzy: in 1988 when the end of the Iran–Iraq War left many suppliers without critical sales; in 1989 when the end of the Cold War undermined domestic procurement; and in 1991 following the overwhelming success of Western equipment in *Operation Desert Storm*.

Despite the serious need for new markets, arms transfers failed to increase as expected. According to the US Arms Control and Disarmament Agency, global arms exports fell from a high of $60bn in 1987 to $22bn in 1993. Virtually all exporting countries altered their policies to facilitate exports, and still arms transfers fell.

There is no unambiguous explanation for the long-term decline of arms exports. The clearest reason is the declining global demand for major new weaponry. Like most manufacturing countries, importing countries are less willing to invest in weapons procurement. The end of the Cold War and the spread of democracy are important elements in this change.

So are technological factors like the growing life-span of weapon systems; with proper care and modernisation they do not wear out as rapidly as they used to. Cascading weapons systems made redundant by the CFE Treaty and unilateral reductions elsewhere have made it easy for many countries to satisfy their needs for reserves relatively inexpensively. And many countries that used to buy weapons profligately now purchase more systematically; even small developing countries are more careful about their military procurement, setting formal requirements and inviting competitive bidding.

Such considerations offer only a partial explanation of what must remain an enigma. Contrary to popular impressions, major arms transfers are increasingly rare. Since the early 1990s, there have been only a handful of new multi-billion-dollar international defence deals: The UK's enormous *al-Yamamah* II deal with Saudi Arabia in 1993, the French sale of GIAT tanks to the United Arab Emirates the same year, and the US sale of F-15s worth $1.74bn to Israel in 1994. The UAE is expected to announce the winner of its fighter replacement competition resulting in a contract worth $6–8bn in 1996. This is expected to be the last major Middle Eastern arms deal for many years to come. East Asia may be the only region left where new orders of major weaponry are routine, but even these orders tend to be restrained, typically involving one or two ships at a time, no more than two dozen aircraft or a handful of artillery pieces. Only by stretching the imagination can the situation be called an arms race.

One of the distinguishing aspects of the contemporary arms trade is the disproportionate significance of smaller deals. The largest arms transfers cause few complaints, if only because they pose few challenges to the geopolitical status quo. Reflecting modernisation rather than expansion, they are increasingly exceptional and tend to reaffirm accepted balances of power. The same cannot be said for some of the smaller arms transfers, such as sales of ballistic missiles or other critical technologies which give the recipient potentially destabilising capabilities. These latter transfers may be far more important strategically, but they mean relatively little for the long-term survival of major defence industries.

In specific cases, arms exports can still be critical to the success of defence manufacturers. For British Aerospace, signing the long-delayed *al-Yamamah* II deal to furnish Saudi Arabia with a wide variety of military equipment and infrastructure worth £20bn ($30.4bn) was critical to its economic recovery. Much smaller exports have been no less important to the survival of Polish, Russian and Slovak manufacturers. But not even large exports guarantee improved fortunes. In an extreme case, a successful bid to supply the UAE with 436 *Leclerc* tanks in 1993 undermined the French ground-equipment manufacturer GIAT. The narrow terms of the deal, compounded by unsuccessful foreign-exchange speculation, left the company with a loss in 1995 of FF11.8bn, against a total income of FF5.3bn that year. The French government has assured GIAT that it will rescue the company, but only after a restructuring plan is in place.

No Innovation, No Survival

The ultimate success of defence manufacturers depends on their domestic market, their only reliable source of long-term contracts. Exports can be of great value, especially when they provide temporary surges in otherwise lean times, but they are too intermittent to sustain large firms. For this reason, the decline in military spending and military procurement will force defence suppliers to react with greater creativity and originality than in the past.

Previously, it was enough for defence firms to rely on excellent engineering to serve their clients and prosper. Assured markets and profits gave them economic security unlike that found almost anywhere else in the economy. Today, with procurement spending falling and military planning highly uncertain, such security is gone. Mergers and consolidations will not be effective alone; they must be followed by sweeping reorganisation. Defence firms must also demonstrate the kind of managerial talents and business foresight previously associated only with civilian enterprises. Even the best management will not eliminate the new risks to the defence business.

Is There A Revolution In Military Affairs?

Although few analysts disagree that a revolutionary shift has occurred in the international security environment since 1989, there is far less consensus on whether a revolution in military affairs (RMA) is under way and, if so, what this might mean for the future of warfare. The present era is roughly analogous to the mid-nineteenth century and the advent of armoured, steam-driven warships. The current period can also be compared to the inter-war years of the 1920s and 1930s. Then, advances in the internal combustion engine, aircraft design and communications were equally available to the UK, France and Germany, but only one of these countries absorbed the operational and organisational import of such advances, exploiting them with the *Blitzkrieg* and decisive victories in 1939–41. Today, the very same information technologies that have opened up closed societies and transformed the global economy are also reshaping the way future wars will be fought. Not recognising the implications of these revolutionary developments for warfare, it is argued, entails grave risks.

Analysts and defence planners are baffled by the challenge of evaluating the current military revolution in the absence of *a priori* knowledge about its precise characteristics and likely evolution. Political elites and interested publics must make informed decisions not just about the aggregate level of defence spending, but also about investment in near-term readiness and force structure versus future military capability. Moreover, the military revolution is not a phenomenon exclusively associated with US superiority. On one level, it is difficult to imagine the United States engaged in major regional conflicts without its allies. These allies, however, may, for fiscal and industrial reasons, have disparate views on whether a military revolution is indeed taking place and how best to approach it. On another level, this revolution is likely to affect regional military balances, not least because potential regional adversaries are likely to pursue selectively new military capabilities that could render Western intervention increasingly difficult and doubtful.

What Is The Revolution in Military Affairs?

The 1991 war in the Gulf only dimly reflected signs of revolutionary changes in warfare. Virtually all the weapons used, however effective, were decades old. Nor were there any dramatic doctrinal or organisational innovations. Indeed, in the area that perhaps best foretells the nature of emerging threats – mobile missiles armed with weapons of mass destruction – the coalition had only marginal success with *Patriot* missile defences and no confirmed kills in the so-called '*Scud* hunt', despite directing nearly 15% of its total air sorties against Iraqi missile launchers and dedicated special forces operations.

Yet there were notable demonstrations of new military technologies in the Gulf War, most profoundly in the various forms of information used and communicated around the battlefield. The coalition had unprecedented access to precise information – from satellite-gathered intelligence, global positioning and meteorological data, maps derived from remote sensing, airborne surveillance radars detecting moving vehicles, to missile-launch detection data. At the same time, Iraq was denied access to information by the coalition's precision strikes, executed swiftly within minutes of the War's start. Information dominance clearly helped to minimise coalition casualties. But new military technology is only one dimension of a revolutionary change in warfare.

A revolution in military affairs occurs when new technologies (internal combustion engines) are incorporated into a militarily significant number of systems (main battle tanks) which are then combined with innovative operational concepts (*Blitzkreig* tactics) and new organisational adaptation (*Panzer* divisions) to produce quantum improvements in military effectiveness. The twentieth century is marked by three military revolutions: mechanised warfare in the 1930s and 1940s; nuclear weapons and ballistic missiles in the 1950s and 1960s; and cybernetics and automated troop control (information technology) beginning in the 1970s and continuing into the twenty-first century.

Historically, organised warfare can be seen as various combinations of attrition (fire power) and manoeuvre (mobility). In this century, the Western Front of the First World War is the classic example of an attrition-dominated conflict, as opposed to the manoeuvre-dominated conflict of the German *Blitzkrieg* campaigns in the Second World War. Whatever the admixture of these two classic ingredients, information about the enemy's strength, location and intentions has always played an important, but secondary military role. Operations were undertaken in the absence of information, and the means for collecting, analysing and disseminating information often failed to pierce the 'fog of war', and in some cases even added to it.

New Tools of War

Many defence experts argue that the emerging military revolution elevates information above both attrition and manoeuvre, allowing them to be applied with heretofore unachievable precision. These experts forecast a military revolution characterised by the ubiquitous employment of microprocessors throughout military force structures, remote sensing technology, advanced data-fusion software, interlinked but physically disparate databases, and high-speed, large-capacity communications networks. The technology fueling this revolution does not destroy anything *per se*, or transport physical objects such as troops and equipment over long distances. Rather, it enables the precise application of force against an enemy's vital centres of gravity, and it supports the assembly and deployment of forces in space and time so as to maximise their operational impact and minimise their own vulnerability.

A key feature of this military revolution is that the underlying technology is driven at least as much by the civilian commercial economy as by government-sponsored military research and development. But this reality has its ominous side. Rather than being relegated to permanent inferiority *vis-à-vis* the developed world, minor powers will have increasing access to capabilities that will directly counter Western military superiority in the twenty-first century. Specifically, the global marketplace will provide the following capabilities to any country, group or even individual with the resources to finance their acquisition:

- Stealthy techniques that significantly lower radar signatures.
- Precision guidance provided by the US Global Positioning System (GPS) or the Russian GLONASS satellites.

- Ship- and land-attack cruise missiles using both of the above technologies.
- Sophisticated anti-ship mines and torpedoes, which can be delivered by advanced diesel submarines.
- Surface-to-air and air-to-air missiles identical to the best in Western inventories.
- Upgrade packages for ageing weapons systems that raise them to the best Western standards.
- High-resolution satellite imagery and sophisticated processing.
- Communications and computing technology equal to, or better than, systems currently deployed with Western force structures.
- Enabling techniques – all dual-use – for weapons of mass destruction (nuclear, biological and chemical).

Even though the concept of the revolution in military affairs is murky and highly contentious at this early stage, the core characteristics of the incipient transformation in modern warfare can be identified. With the level of situation awareness potentially available through advanced information technology, firepower can be brought to bear simultaneously throughout a theatre of operations, denying an enemy the ability to recover from the kind of sequential attacks that formerly characterised attrition-dominated warfare. With respect to manoeuvre-dominated operations, the linear battlefield with its front, rear and flanks will dissolve into the non-linear battlespace – wherein small, highly mobile, and extremely lethal forces can create operational and strategic-level effects against larger, traditionally equipped and led opponents. Sophisticated forms of information warfare and long-range precision weaponry even raise the prospect of inflicting strategic damage on a country's national assets without using weapons of mass destruction.

Competing Schools of Thought

The existence of a revolution based on information technology, whose implications can only be dimly perceived, will become a source of increasing debate in Western military establishments as those implications challenge entrenched thinking and associated economic interests. The current state of the debate in the US is illustrative. Shrinking defence budgets and international systemic uncertainty have brought the debate into focus for two competing groups. The outcome of the resource allocation and doctrinal debate between these two factions will profoundly affect US and Western security policy and force structure for the foreseeable future.

Platform-oriented Traditionalists

In the US, one faction is composed of sceptical incrementalists seeking to build upon the enormous sums already invested in weapons platforms (armour, aircraft and capital ships). The current generation of platforms, it argues, comprises extremely flexible systems that can be upgraded with advanced technology to improve their lethality and are, therefore, a cost-effective response to the evolving international environment. Not surprisingly, the threats this faction perceives support such a resource allocation. It hinges critically on focusing the bulk of US defence resources on preparing for two nearly simultaneous major regional contingencies (e.g., against Iraq or Iran in the Middle East, and North Korea in East Asia). In addition, several lesser regional contingencies range from a large-scale peacekeeping/peacemaking operation (Bosnia) to a small-scale peacekeeping/humanitarian one (the Former Yugoslav Republic of Macedonia or Rwanda).

The platform advocates, who represent key states and US Congressional districts with work-forces entrenched in the platform and subsidiary businesses, are led by senior military officers and legislators who look at a 5–10-year timeframe. Deep-rooted bureaucratic advocacy for platforms also comes from multiple layers of middle-level officials representing the civil service, armed forces, government R&D organisations, and industry, who share the responsibility of integrating platform subsystems distributed broadly throughout the country.

Information-oriented Modernists

Proponents of the view that a military revolution is looming tend to look forward 20–40 years and envisage a threat environment divided into three categories, roughly the short, mid- and long terms. The first is a deadlier strain of rogue states than the West has faced before, one that forswears Cold War-era, armour-heavy conventional operations in favour of unconventional methods that impose penalties costly enough to deter or limit Western action. Relatively cheap (compared to aircraft) ballistic and cruise missiles armed with weapons of mass destruction (foremost, increasingly available biological agents) will become the delivery means of choice, while small-unit operations and environmental (or 'dirty') warfare involving water supplies and industrial plants will gain in importance. The regional opposition of the future need not risk open hostilities with a superpower, but will be able to conduct information operations against vulnerable Western governmental and commercial infrastructure either regionally or globally.

The second threat these advocates see comes from extra-national organisations that could directly endanger the industrial world's vital interests. Three immediate and topical examples include Islamic radicals (oil supply); Russian 'mafia' (proliferation of weapons of mass destruction); and Latin/Asian drug cartels (domestic societal and regional instability).

The third aspect of the military revolution threat involves the emergence of a peer competitor. While the specific identity of the next superpower is difficult to predict, RMA advocates believe that a rival power bloc to the West is likely to emerge in the longer term. Currently, the most likely region to rival the West is Asia, but the central belief is that economic competition will fuel a potential adversary's military technology development more efficiently than in any previous period in history given the dual-use nature of modern information technology.

These proponents come from the armed forces, government and industry, but are very different from the platform proponents. They are led by a small group of senior military personnel who take a longer-term and less service-parochial view of national security. They have natural allies in the legislative branch, some of whom are so-called 'cheap hawks', who champion more affordable approaches to meeting longer-term defence needs. Within the growing industry of commercial information technology there are also strong proponents for change. They see a revolution in military affairs as providing them with new market opportunities, especially as military procurement rules become more open to commercial competition. An increasing number of analysts has also begun to ally itself with other supporters throughout the military services, government and think tanks. Although expanding, their size and influence still pale in comparison with platform-oriented traditionalists.

Whither the US Debate?

In the struggle to shape US military thinking, platform-oriented traditionalists currently dominate what stands for long-term military planning. This group has successfully portrayed the Gulf War as the first post-modern conflict rather than the last platform war. The importance of this view can scarcely be exaggerated since it is shaping the modernisation of the US force structure with respect to timing and composition and, perhaps more importantly, to miliary revolution proponents, depressing defence-related R&D in both relative and absolute terms. Current US military R&D constitutes $35bn out of a $250bn budget, down from a peak of $45bn out of $343bn (current dollars) in 1986. It will shrink to $24bn out of a $220bn defence budget by 1999.

If the late 1990s signal the beginning of a genuine change in military affairs, then force-structure recapitalisation should be delayed until it becomes clear what the capital ship of the new warform is likely to be. The platform advocates argued that major elements of the force structure are reaching obsolescence, while their opponents are more inclined to see ageing platforms as candidates for extinction rather than replacement. They feel it would be best to emulate the start of the platform era, when the services procured small numbers of 60 types of fighter aircraft and built very few capital ships in order to study the technology before deciding on the optimum force structure. Although major hardware programmes are not expected to be entirely discontinued, proponents of the military revolution foresee them as assuming more of a secondary or supporting role to advanced technology.

Recapitalisation may be more a political than a military necessity, and platforms can serve as bridges to a force structure better adapted to the proposed military revolution through information technology upgrades – at least for traditional threats. Although equipment modification cannot be deferred indefinitely, RMA proponents argue that even an inventory of obsolescent US weapons so greatly exceeds those of potential adversaries that the US can afford to wait and see what new weapon requirements a possible military revolution suggests. Thus, they are most concerned with the long-term decline in the military research-and-development budget.

This debate is likely to assume increasing prominence during the 1996 US presidential election year. Not least of the issues expected to be debated will be whether the US can afford the force structure and readiness costs of preparing to fight two nearly simultaneous major regional contingencies, which forms the heart of the Clinton administration's 1993 *Bottom-Up Review*. Equally, if not more, important is the question of the extent to which presidential leadership will be exerted within and among the military services. Until the military more vigorously embraces new organisational and operational concepts, many of which involve truly integrated multi-service operations, it seems likely that any military revolution based on US leadership will be a long, slow evolutionary process.

Is the Military Revolution a World-wide Phenomenon?

If the United States has failed to embrace the current revolution as aggressively as its advocates would have liked, does it really matter, given that no comparable military power threatens US military superiority? Those pushing for change would suggest that there are compelling reasons for concern over which changes the United States does or does not adopt. If the US chooses the new technologies with fervour, potential competitors are more likely to be deterred from trying to keep pace. Conversely, US failure to pursue RMA changes vigorously could have severe consequences in both the near and longer term. To assume that the US will have military superiority indefinitely is shortsighted because the information-based revolution in technology is commercially, not governmentally, driven. The technologies are broadly available to any nation willing to pay for them and integrate them into its military systems. They could facilitate surprisingly rapid doctrinal and organisational changes, perhaps not on a scale comparable to US advances, but large enough to change regional military balances or affect great-power relationships. It is thus useful briefly to review the position of other key nations or regions with regard to the revolution in military affairs.

Russia: Past as Prologue?

Ironically, Soviet military practitioners were the first to embrace the notion of an emerging revolution in military affairs, albeit largely rhetorically. In the 1970s, Soviet military writers began drawing attention to a new revolution caused by innovative military electronics, including computers, sensors and communications systems, tied to increasingly long-range conventional weapons. These new capabilities, they argued, produced a qualitative change in conventional warfare, making non-nuclear means virtually equivalent to very low-yield nuclear weapons in military effectiveness.

Because long-range, highly accurate firepower no longer needed to be concentrated, like artillery, Soviet military theoreticians foresaw the need to rethink the relationship between offence and defence. As weapon ranges increased and tactical and operational manoeuvre became standard on the battlefield, defence could rapidly cross over to offence. Despite severe shortcomings in the military electronics that were then, and are now increasingly, available to Western military forces, by the mid-1980s the Soviet military began to experiment with new operational concepts ('manoeuvre by fire') and even with related changes in force structure (Operational Manoeuvre Groups).

It seems clear, in retrospect, that early Soviet interest was driven in part by growing concern that technology base shortcomings would prevent it from keeping up with Western military developments in the twenty-first century. The most vocal proponent for change was Marshal Nikolai Ogarkov, Chief of the Soviet General Staff from 1977–84. Ogarkov strongly advocated much greater emphasis on the new military technologies supporting long-range precision strikes, but ran headlong into traditionalists, primarily in the ground forces, who favoured a more evolutionary approach oriented around tank warfare. Although there were probably a variety of reasons for Ogarkov's surprise removal as Chief of the General Staff, his unrelenting pressure to adopt revolutionary changes – including calls for budget increases to finance high-technology innovation – undoubtedly helped assure his departure. Yet even his successor, Marshal Sergei Akhromeyev, continued, if more delicately, to challenge the military not to plan for the last war.

After half a decade of economic disintegration associated with the break-up of the Soviet Union, Russia's military now faces enormous, if not insurmountable, problems. Simply maintaining the semblance of coherently structured armed forces, even without exploiting a revolution in military affairs, is a daunting task. Rebuilding Russian forces to a level competitive with US military forces will require not only sustained economic growth and political stability, but also a fundamental transformation of military industries and the absorption and assimilation of new information technologies. Given Russia's currently tenuous circumstances, additional erosion in military capability is likely to precede any ability to realise the implications of the military revolution, which is likely to unfold over decades rather than in the immediate future.

China: Sleeping Giant?

China holds the key to regional stability in Asia, and may become a peer competitor with the US for global influence in the next century. The extent to which it adapts its military is thus of considerable significance. The first step for any country is to acknowledge the need for radical change. Chinese military strategists have gone further and are now devoting increasing attention to the problems in store. While the path ahead might seem clear to many Chinese strategists, its implementation will be constrained not least by the complexity and cost of developing an indigenous manufacturing infrastructure for high-technology weapons. For this reason, the RMA is unlikely to figure quickly in China's military structure. But determined military reform coupled with sustained investment over at least two decades could conceivably transform China into a formidable military power, certainly regionally, if not globally.

For the past decade, China has been trying to move beyond Mao's 'People's War' strategy, which emphasised quantity over quality, to a new strategy oriented towards incorporating high technology into increasingly sophisticated forces for limited regional conflict. Chinese strategists seem to have been heavily influenced by the 1991 Gulf War, which underscored the growing importance of electronic warfare, long-range precision strike, advanced reconnaissance, and much-improved command, control, communications and intelligence. All these are areas of severe infrastructure weakness. To help compensate, China has been buying from abroad (notably Russian systems, including the Su-27 advanced fighter and IL-76 transport aircraft, and the SA-10 air-defence system) and has expressed interest in co-production and co-development programmes, including one with Israel. Still, the path ahead is arduous: China faces block aircraft obsolescence, and the costs involved in matching its high-tech weapon desires are huge. In the end, China's approach is likely to be highly selective; a mixture of older, upgraded systems, and advanced-technology weapons capable of long-range projection (for example, land-attack cruise missiles instead of more costly manned aircraft).

Europe: Struggling to Adapt to New Realities

While US proponents of the military revolution might be small in size, they are vocal in their advocacy. It is hard to find much of a constituency in Europe, vocal or otherwise. This should come as no surprise, for the United States' NATO partners are more caught up at the moment in the struggle to maintain a defence-industrial base geared to satisfying a shrinking internal need, while at the same time competing globally with an increasingly aggressive US defence industry. Trying to fit the military revolution into European defence consolidation seems to have become lost in a flurry of mergers and acquisitions.

Driven in part by fear of what US defence consolidation might mean competitively, defence industries in France, the UK, Germany and Italy seem destined to consolidate or to form joint decision-making alliances. Yet, from the standpoint of the military revolution, it is not at all clear that either US or European defence consolidation will result in better industrial capability to produce information-age technologies or cheap, large distributed systems. To effect real change requires the widespread adoption of commercial practices within consolidated defence industries, as well as assured access to commercial R&D. Thus, to the extent that existing defence consolidation is predicated on producing traditional, platform-based solutions, it may simply be further delaying changes in military institutions consistent with a military revolution.

Yet there are some signs of movement in Europe that could hasten more changes in the longer run. Like the United States, the UK and Germany have cut their defence spending by about one-third since 1991. France's decision to abolish its draft and to modernise its force structure has created an opportunity for real change. Coupled with its increased investment in intelligence-gathering satellites (*Hélios*) and long-range precision strike means (*Apache* cruise missile), France's February 1996 decision to streamline its armed forces and make them more suitable for high-technology warfare (rapid out-of-area force projection) certainly suggests a more appropriate model for the emerging threat environment.

Other Military Powers: Introducing Selective Responses?

Nations with relatively clear strategic objectives, but substantially less capital investment in military platforms at risk from a revolution in military affairs, could choose to move away from huge investments in vulnerable platforms and invest more wisely, if only selectively, in technologies and systems more relevant to the new thinking. Such countries could well include Iran, Libya, Taiwan and other Pacific Rim nations anxious to acquire advanced military capabilities, but not necessarily large and expensive force levels. Such countries can be expected to pursue an unconventional path to regional power that would not depend on traditional platform-oriented military solutions.

Instead, these emerging military powers are likely to be more attracted to numerous cheap delivery means (largely cruise and ballistic missiles instead of more expensive manned aircraft), commercial overhead imagery for targeting GPS-guided weapons, penalty-imposing countermeasures to projection forces (sea mines and diesel submarines), and small-unit operations designed to face larger, traditional manoeuvre forces.

The Road Ahead

There seems little doubt that the pace of technological change makes transition to new forms and methods of warfare inevitable. How quickly or evenly that transition might unfold, however, is not clear. The pace of change is not just the product of technology absorption precipitating doctrinal and organisational adaptation. Just as important is the level of military competition between states and the relative degree of stability in the rapidly changing international environment.

Regarding future stability, there are reasons for both optimism and pessimism. On the optimistic side, there is general tranquility amongst the major powers which, along with most other industrialised nations, share democratic goals and market economies. In addition, the United States remains the sole unchallenged military superpower. Yet, on the more pessimistic side are potential instability in the two large transition states, Russia and China; the increasing propensity for ethnic conflict and state fragmentation; fragile Cold War alliance systems; and the prospects for unbridled proliferation, not least because of the greater availability of advanced technology. At the moment, most major and minor states appear to lean towards the more optimistic outlook.

Any military revolution that does unfold will probably do so unevenly. The US would seem to have distinct advantages over its competition. Given US leadership in the relevant commercial sectors, as well as in overall economic performance and defence resource allocation, it would appear natural for the United States to continue exploiting its early lead in absorbing new information technologies. But the US has demonstrated less notable progress in doctrinal and organisational change, without which its military transition will be fragmentary and prolonged. Moreover, history shows that relative economic and industrial performance does not always determine the outcomes of military competitions. Despite substantially smaller gross national products than the US, Germany and Japan took the lead in the inter-war years in implementing revolutionary changes, in manoeuvre warfare and naval aviation respectively. Necessity is frequently the driving force for change – in Germany's case, treaty-imposed manpower restrictions led to covert experimentation with new forms of tank warfare. Such examples raise fears in some quarters that the leisurely pace at which the US is embracing the new concepts will allow a future challenge from a new peer competitor for great-power status.

Because allies are likely to fight alongside US forces, whether NATO countries embrace the emerging military revolution in a coordinated or *ad hoc* fashion will have a telling effect on alliance military options. Although the direction of the NATO-led Bosnian peacekeeping operation (IFOR) remains uncertain, there have already been unprecedented signs of military coordination. US JSTARS reconnaissance aircraft now feed radar information directly to French forces, while both French and UK forces exchange intelligence on activities within US zones of responsibility. However small, such efforts reflect the enormous potential inherent in skilfully manipulating electronically transmitted information – the core element in the unfolding military revolution. But there is much to overcome, above all a history of failed attempts at US–European equipment development and intense competition for shrinking international arms sales. Whether short-term economic gain or long-term mutual security interests prevail will in part be seen in the outcome of such nascent RMA-like efforts as the joint US, French and German programme to develop a medium-range extended air-defence system, called MEADS.

Two additional factors are critical in evaluating the military future. The first bears on a major challenge facing twenty-first-century military forces: safely projecting military power against regional adversaries armed with weapons of mass destruction. Given the West's growing intolerance to accept casualties – reinforced greatly, if erroneously, by *Operation Desert Storm* – how can vulnerable ground, air and naval forces be deployed within harm's way? Exacerbating this growing vulnerability is the second concern that future Western nuclear threats will no longer have the same deterrent effect they did on Saddam Hussein in 1991, particularly in light of the diminished status that nuclear weapons now have within US military circles. The expectation of military revolution proponents is that conventional improvements in ubiquitous battlefield intelligence and stand-off precision strike will eventually compensate for current conventional and nuclear force weaknesses. The challenge is to implement these changes before the capacity for intervention and, more importantly, the credible threat of intervention becomes a relic of the past.

The Role Of Non-Lethal Weapons

The debate over non-lethal weapons in some respects reflects the modern challenges of exercising military force. National policy-makers and military leaders, particularly in major military powers, often face a political dilemma in employing their armed forces. They are expected to make effective use of their armed forces in dealing with security threats or supporting peace operations, but to do so without incurring or inflicting substantial casualties – particularly among non-combatants.

The renewed interest in non-lethal-weapon technologies stems from the new challenges that the United States and other major military powers face in conducting peace and other low-intensity operations given such demanding political imperatives. Military commanders, as well as their political leaders, seek expanded options for dealing with complicated local situations, particularly in failed states where using lethal force is likely to escalate the level of conflict. Near-instantaneous transmissions by the international media of graphic images of death and mutilation from the war zone, which can profoundly shape national and international public opinion, only add to the interest in non-lethal options. Thus, some political and military leaders are seeking options that fall between doing nothing, and resorting to lethal force in volatile situations. Advocates of non-lethal weapons argue that new technologies offer intermediate options to employing full military force. This has generated debate over whether non-lethal weaponry provides a unique means for conducting lower-risk military operations in armed conflicts, or whether such weapons are likely to be limited at best to relatively narrow applications.

Types of Non-Lethal Technologies

There are two basic types of non-lethal technologies, anti-personnel and anti-material. The former subdue individuals and groups, while the latter deny them the use of vital equipment, including vehicles and even weapon systems. Table 1 shows a range of anti-personnel and anti-material technologies with potential military applications as non-lethal weapons. Such non-lethal weapons can be employed by armed forces to incapacitate threatening groups temporarily, or to disable their equipment, in a way that limits the level of violence and the risk of casualties to non-combatants. Of course, even these so-called non-lethal weapons can produce lethal results depending on how they are used and the physical condition of individuals subjected to their effects. Hence, the term 'less-than-lethal' has gained greater currency in describing weapons that are not intended to kill or permanently injure human beings.

Anti-personnel weapons include a broad range of non-lethal technologies. Some technologies are variations of the more familiar crowd-dispersal weapons, such as guns and grenades that spray rubber pellets, bean bags, foam or wooden baton rounds. More advanced anti-personnel technologies would disable individuals using either sticky foam dispensed from a high-pressure gun system or a system that seeks to contain disorderly groups by creating a foam barrier containing a tear gas agent. Another technology for warding off a threatening crowd uses advanced infrasound generators to emit very low-frequency sound waves strong enough to disable nearby individuals temporarily.

Table 1 *Non-Lethal Weapons*

Type	Target	Description
Acoustics	personnel	Sound generator causes severe pain and disorientation
Chemicals	personnel	family of chemical agents that incapacitate personnel or even equipment
Immobilisers	personnel or vehicle	sticky foam, super-adhesives and anti-traction compounds that greatly inhibit the movement of personnel or vehicles
Light sources	personnel or equipment	low-energy lasers that can blind personnel and disable electro-optical equipment
Microwaves	equipment	high-powered microwaves to disable electronic systems
Power disrupters	equipment	various types of chemical agents or conductive ribbons that can disable vehicle engines or power sources, or biological agents that can degrade vehicle fuels

Note: Other anti-personnel systems include obscurants (e.g., smoke), entangling nets and devices, and a wide variety of projectile weapons. Other anti-material technologies include chemicals that can solidify fuels or cause critical materials to become brittle and disintegrate, and even computer viruses.

Sources: Office of Technology Assessment, *Improving the Prospects for Future International Peace Operations* (Washington DC: Office of Technology Assessment, September 1995); and *Aviation Week and Space Technology*, 16 October 1995, p. 50.

Many anti-personnel weapons stem from technologies devised for civilian applications, particularly for US law-enforcement agencies. In recent years major police departments and riot-control agencies have absorbed a broad range of non-lethal technologies and techniques. Hence, the non-lethal technologies suited to crowd control and dispersal have already been proven in the field. But many technologies outlined in the Table, particularly the anti-material weapons, have yet to proceed beyond field testing under very controlled conditions. Indeed, several are still only in the design stage.

A wide range of advanced non-lethal technologies is potentially available for use against vehicles and weapon systems. One class of technologies seeks to immobilise vehicles, either by using powerful adhesive to fix their location or anti-traction solutions that make roads and airstrips unusable by negating their friction qualities. Other concepts envisage using chemical agents to neutralise vehicles and weapons by degrading their rubber components (e.g., tyres, hoses and insulation), or even using microbes to degrade fuel and explosives by rapidly accelerating biodeterioration.

Some anti-material technologies have broader military implications beyond supporting non-lethal operations. For example, high-powered microwaves, which can destroy critical electronic components of modern weapon systems and vehicles, might play a potentially important role in supporting regular combat operations similar to existing electronic warfare capabilities. Immobilisers could be used in combat operations in ways that make enemy troops more vulnerable to capture, wounding or being killed. Finally, computer viruses, which can be deliberately injected into an adversary's computing and communication systems, offer a form of information warfare that can disrupt not only military forces, but also national economic, energy and transport networks.

The Political–Military Context for Non-Lethal Weaponry

At least conceptually, both types of non-lethal weapon technologies are relevant to three basic kinds of military mission. At the lower end of the spectrum, non-lethal weapons have been employed by armed forces assigned to certain non-combat missions, such as crowd control and dispersal. Similar missions include defending embassies from rioters or conducting hostage-rescue operations. The purpose in these cases of using non-lethal weapons, such as wooden baton rounds fired from guns, is to minimise casualties among non-combatants and to avoid escalating an already volatile situation. Although new non-lethal-weapon technologies, such as foam barriers and acoustic generators, might offer greater operational effectiveness over earlier capabilities, they are unlikely to produce new military options.

In comparison, non-lethal weapons have also traditionally played a significant role at the higher end of the military force spectrum by supporting conventional battlefield operations. For example, electronic warfare systems and reconnaissance capabilities perform a critical role in modern combat operations, even though they do not directly kill or wound military troops. Certain advanced non-lethal-weapon technologies, such as high-powered microwave systems, might figure prominently in high-intensity combat operations. If proven to be technically feasible and operationally effective, such non-lethal weapons could become standard equipment in the force structures of modern military establishments. The military value of such non-lethal weapon technology, however, stems largely from their ability to serve as force multipliers for lethal forms of military force, rather than from the contribution they make to non-lethal operations.

The most uncertain aspect of non-lethal weapons, however, is whether they offer new options for conducting military operations, particularly when a premium is placed on minimising casualties. Peace operations or humanitarian missions in failed states, where foreign troops risk bodily harm in performing their assigned duties, may cause a local clash to escalate rapidly catching non-combatants in the crossfire. Some advocates of non-lethal weapons contend that new technologies and tactics provide national policy-makers and military commanders with intermediate choices between doing nothing or using lethal force in volatile situations, such as subduing armed opponents who use non-combatants as human shields. Other problematic peacekeeping missions include preventing the theft or seizure of valuable commodities (e.g., relief supplies); segregating and dis-arming local combatants; and dispersing threatening or obstructive crowds.

Prospects for Non-Lethal Weapons

The controversial proposition that improved non-lethal weapons can offer new, lower-risk options for applying military force in armed conflicts raises several questions. Do military forces employing non-lethal weapons have the advantage in a confrontation, or will the opponent's responses and possible countermeasures make any military edge short-lived at best? Are military establishments willing to adopt doctrines that make substan-tial use of non-lethal weapons, as well as provide the necessary funding and training for such non-traditional weapon systems? And will adopting non-lethal weapons be inhibited by existing arms-control restrictions and concerns that some technologies create new forms of inhumane weapons that are better banned than encouraged? The answers to these questions will determine what roles non-lethal weapons are most likely to play in future conflicts.

Combat Utility

The willingness of political leaders and military commanders to consider non-lethal weapons for expanded roles in armed conflicts will depend on their perceived utility in combat situations. In theory, non-lethal weapons are well suited to support peacekeeping and humanitarian operations in failed states where any use of force that fails to discriminate between non-combatants and those instigating violence and disorder will only compli-cate the political–military situation. But the benefits of using non-lethal weapons could be negated in situations where determined and organised adversaries are threatening the peace-operation forces.

Employing non-lethal weapons might even add to the general confu-sion by giving policy-makers the false hope that non-lethal technologies will somehow achieve desirable results in a situation where peacekeeping forces must otherwise depend on some level of consent from the local opposition to perform their missions. Instead of accepting defeat in a particular confrontation, organised opponents might choose to escalate the conflict by responding with deadly force to the peacekeepers' use of non-lethal weapons. How long will peacekeeping troops refrain from returning lethal fire when they run the risk of being killed and wounded during an attack? Furthermore, military commanders are likely to fear that failing to undertake a decisive (i.e., lethal) response will only encourage their opponents to launch even bolder attacks against their forces.

Countermeasures could pose another potential challenge to the military utility of non-lethal weapons. The military advantages offered by such weapons could be substantially diminished for peace operations if the opposition can acquire technical countermeasures or, more likely, adopt offsetting tactics that significantly degrade the effectiveness of particular non-lethal weapons. Hence, depending on the specific circumstances, non-lethal weapons might give peacekeeping forces no more than an initial advantage. Sustaining this technological advantage could be difficult over a protracted operation, especially if the opposition acquires technical countermeasures from foreign suppliers or devises its own expedient tactics for circumventing the effects of non-lethal weapons.

Military Doctrine and Force Structure

How military leaders view the combat utility of non-lethal weapons will also have an important bearing on whether these non-traditional weapon systems ever become integral to the military doctrines and standard equipment holdings of regular armed forces. Defence planners recognise that certain non-lethal weapons have an important role to play in performing very specialised tasks, such as hostage-rescue missions, as well as serving as combat support systems. Many military professionals are sceptical for several reasons about relying on non-lethal weapons to perform missions that have traditionally been accomplished by combat forces; such weapons do not provide the same degree of confidence in achieving battlefield outcomes as lethal weapons. Conventional arms, such as artillery and tanks, can generate 'hard kills' by causing casualties and destroying equipment. But non-lethal weapons generate more ambiguous results; they produce what some call 'soft' or 'mission' kills based on temporarily incapacitating the other side's personnel or equipment.

The effectiveness of most new non-lethal technologies has yet to be proven in combat or even systematically tested under realistic operational conditions. For some technologies, such as barrier defences, weather conditions can have a major impact on their effectiveness. And even if advanced technologies, such as immobilisers based on anti-traction substances, prove to be technically sound, military planners must develop tactics enabling their forces to make use of such weapons without having their own operational flexibility substantially degraded during a confrontation.

Some military leaders are concerned about how non-lethal weapons might affect their warfighting doctrines. They fear that adopting non-lethal weapons could blur the military's traditional doctrinal emphasis on applying lethal force effectively to accomplish military missions. Even in peacekeeping and humanitarian support missions, the ability to threaten lethal force with credibility against opponents is seen as vital to the success of such operations.

Military planners must deal with the thorny organisational and budgetary issues posed by non-lethal weapons. Most non-lethal weapons require troops to undergo specialised training and create unique logistical requirements. Hence, should special purpose units be created to employ these weapons, or should non-lethal weapons be integrated into the equipment holdings and training plans of all regular combat units? Will the procurement and training costs of fielding non-lethal-weapon capabilities occur at the expense of traditional combat arms? Such concerns encourage many military planners to view non-lethal weapons mainly as 'nice-to-have' supplements to their primary combat armaments only as long as extra funding is provided for these special capabilities.

These difficult issues help explain why adopting non-lethal weapons is receiving a mixed response from some countries' military establishments. Perhaps the greatest enthusiasm for exploring a broad range of non-lethal-weapon options currently exists within the US Department of Defense. The acceptability of non-lethal weapons was enhanced when the forces covering the withdrawal of the United Nations Operation in Somalia (UNOSOM) in 1995 included some US military units specially equipped and trained to use non-lethal anti-personnel technologies (e.g., barrier foam and sticky foam systems), as well as various projectile weapons for dispersing crowds. Although non-lethal technologies were not actually used during the Somalia withdrawal, the willingness of senior US Marine Corps leaders to adopt these non-lethal weapon systems has added greatly to their credibility.

More recently, US planners have adopted an official Department of Defense policy on developing the necessary non-lethal weapons and doctrine to reinforce deterrence and expand the range of options available to commanders in situations where lethal force is not required. The Pentagon is also preparing long-range plans entailing substantial spending on new non-lethal-weapon technologies. Willingness to try the weapons, however, does not guarantee their success against the more established weapon-procurement priorities of the military services in their battle for scarce budget resources. In comparison, there appears to be markedly less interest among other leading military establishments to pursue a broad array of non-lethal-weapon capabilities.

Political Concerns

Several non-lethal-weapon technologies have raised political and arms-control concerns. First, some of these weapons have been criticised as inhumane. Weapons are generally perceived as inhumane if they cause excessive injury to humans or produce relatively indiscriminate effects that could pose a high risk to non-combatants. Blinding lasers, a key non-lethal-weapon technology that could destroy an enemy's optical and electronic sensors, were identified by government experts at the UN Conference in Vienna on 25 September to 13 October to review the 1980 Convention on Conventional Weapons (CCW) as potentially inhumane weapons that could inflict particularly severe eyesight damage to personnel, including those operating military optical sensors. The Conference report urged nations to forego any laser systems specifically intended to blind personnel. While some countries, such as the United States, have agreed not to develop such lasers, they remain committed to fielding laser systems intended for targeting and range-finding purposes that could incidentally cause human blindness. Other non-lethal technologies, such as high-powered acoustic weapons, which would incapacitate people by inducing disorientation and severe physical reactions, are likely to raise similar concerns about their harmful effects on humans.

Second, some non-lethal-weapon technologies are considered potential violations of key arms-control agreements. Although no arms-control agreement specifically restricts non-lethal weapons, some of the chemical and biological agents being explored as non-lethal weapons raise issues related to the 1993 Chemical Weapons (CW) Convention and the 1972 Biological Weapons (BW) Convention. Certain types of chemical agents (e.g., super-adhesives, anti-traction agents and embrittlement compounds) that could inhibit the movement of personnel and even vehicles pose difficult compliance questions for the CW Treaty because some experts believe that they fall within the definition of chemical weapons. Similarly, certain biological agents that incapacitate personnel, as well as microbes intended to degrade fuel and sensitive equipment components, are criticised as potentially violating the BW Convention.

Finally, some experts have indicated that a global diffusion of these advanced technologies could occur as they spread beyond just the leading military powers. Given the growing connections between technologies that nations pursue for civilian and military applications, there are fewer obstacles to the spread of non-lethal-weapon technologies, either to other nations or even to determined groups. Similarly, certain technologies, such as high-powered microwave weapons or blinding lasers, could proliferate rapidly if the leading military establishments signal a strong interest in them both as non-lethal weapons and as systems to support combat forces. Ultimately, such considerations might dampen the interest of Western policy-makers for procuring and using non-lethal weapons.

Uncertain Prospects

Despite the ingenuity of many non-lethal-weapon technologies, the decision to employ military forces remains risky for national policy-makers and their military commanders, particularly in situations where determined political opponents exist. Nonetheless, new non-lethal-weapon concepts and technologies promise that a broader repertoire of military capabilities will be increasingly available to armed forces around the world. There is no doubt that new non-lethal weapons, such as immobilising devices, will improve the capability for dealing with special situations, such as protecting an embassy against attacks by a hostile mob or defending facilities and equipment parks from forays by thieves or armed marauders.

There is considerably more doubt that non-lethal weapons will be sufficiently integrated into military doctrines and force structures to offer a viable alternative to lethal force when combat troops believe they are under threat. Any serious movement towards fully exploiting the advantages of non-lethal weapons will probably require some successful – or at least fairly promising – demonstrations of their effectiveness. Without some realistic demonstrations of how non-lethal weapons can effectively resolve a life-threatening situation, military commanders and political leaders will be reluctant to lessen their traditional reliance on lethal force to any significant degree.

The Problem Of Combat Reluctance

A new and rather pessimistic view of international affairs is developing; one that focuses on a perceived lack of real leadership. The promise that either the United Nations or rapidly formed *ad hoc* coalitions would be able to stamp out regional and inter-ethnic conflict withered almost as soon as it flowered. Instead of order there is disorder; instead of containment there is withdrawal. Although the United Nations is committed to peacekeeping operations on a scale never seen before, in many ways it is the limits of outside military intervention that have been demonstrated in Somalia, Haiti and, above all, in Bosnia. Multinational institutions have singularly failed to cope with the horrors of ethnic cleansing, be it in Rwanda or the former Yugoslavia. The United States is now the only remaining superpower, the only country with the global reach and logistical back-up to make military intervention work.

Yet, according to this prevailing analysis of the international system, the United States has become isolationist, preferring to focus on its domestic agenda. The hesitancy of President Bill Clinton's foreign policy is said to owe much to this changing mood. On Capitol Hill and among the public at large there is no appetite for foreign adventures. As so often, it is the military in the Pentagon who have been the most cautious about projecting force overseas for often uncertain ends. In the United States there is said to be little stomach for the casualties that a role as the world's policeman might bring. In short, a phenomenon of what might be termed 'combat reluctance' has developed. A superpower is unwilling to flex its military muscle; a President cautiously assesses the political climate before putting his servicemen and women in harms way. The contemporary United States has indeed been described as 'a self-deterred power' – a country preventing itself from using military force.

To an extent and in different ways, this phenomenon of combat reluctance afflicts most of the major Western democracies. Some, like Germany and Japan, carry with them historical and constitutional baggage that makes any military role beyond their frontiers controversial at best. Both these states are slowly becoming more involved in military aspects of international diplomacy, but neither has reached the position of the UK and France which dispatch troops with much greater ease and much less public debate.

Indeed, it is these two middle-ranking military powers, albeit with lower-ranked economic muscle, that have seized upon peacekeeping in the post-Cold War era as a means of maintaining their status at the top of the international diplomacy league. Both, to use the phrase of former UK Foreign Secretary Douglas Hurd, are 'punching above their weight'. But the scale of UK and French involvement in peacekeeping should not be taken as an indication that London or Paris are especially enthusiastic about overseas military commitments. Far from it. They have joined UN forces that have been sent on clearly limited missions with sometimes restrictive rules of engagement. The willingness to deploy troops to a zone of conflict should not be confused with the political will to employ them aggressively.

Debating the Use of Force

In the United States, the concept of combat reluctance appears to be three-cornered. There is the initial reluctance of the military to use any force except under the most favourable of circumstances; there is the heightened concern of both Congress and the US public about casualties; and there is the hesitation of the White House, where a President, new to the world of foreign policy, attempted to navigate a course through the uncertain waters of a chaotic world. Bill Clinton has had the daunting task of establishing both his own foreign-policy credentials and maintaining US leadership, while reconciling the competing pressures from reluctant centurions, a sometimes hostile Congress, a sceptical public and a foreign-policy establishment that sometimes tends towards a more active role for the military than does the Pentagon.

Of course Washington has been here before. Today's debate is very much a reflection of themes last raised in the mid-1980s – and the circumstances are not altogether different. In November 1984, then Secretary of Defense Caspar Weinberger established a set of criteria for the use of force, to which Secretary of State George Shultz responded with his own, dissenting thesis. The so-called Weinberger doctrine established six tests to be applied before committing troops to combat.

- Forces should not be committed unless the action is vital to the US national interest or that of its allies.
- Forces should be committed wholeheartedly with the intention of winning, or they should not be committed at all.
- Forces should be committed with clearly defined political and military objectives.
- The use of force should be a last resort.
- The relationship between objectives and the force committed should be continually reassessed and adjusted if necessary.
- With the Vietnam experience very much in mind, before committing forces abroad there should be some reasonable assurance of public and Congressional support.

Weinberger's approach was influenced by the disaster suffered by the US Marines in Lebanon. In October 1983, some 240 US Marines were killed when a truck-bomb was driven into their barracks in Beirut. In retrospect, their mission seemed critically ill-defined. Indeed, the Beirut episode had a similar impact to the more recent incident in Somalia in October 1993 when 18 US soldiers died in an ambush and three helicopters were shot down. Imprecision about goals and means in Somalia had created a fatal 'mission creep'; avoiding this has become almost a fixation among US military planners in Bosnia.

In response to the Weinberger doctrine, Shultz presented his own, less restrictive views about the use of force. In particular, he argued that the US must countenance military action in a wider range of contingencies and not just when its vital interests were at stake. He argued that 'for the world's leading democracy the task is not only immediate self-preservation, but our responsibility as a protector of international peace, on whom many other countries rely for their security'.

The current debate is centred between the restrictive views of Weinberger and the more expansive vision of Shultz. Officials in both the Bush and Clinton administrations have generally accepted that force can be used when problems fall short of vital interests. Secretary of Defense William Perry emphasised in his Forrestal Lecture to the US Naval Academy in April 1995 that some important national interests may not quite attain the status of vital interests, but may require the dispatch of force. Weinberger's emphasis on winning and matching force levels to clear political goals was reflected in General Colin Powell's September 1993 speech as out-going Chairman of the Joint Chiefs of Staff, when he emphasised the need to use decisive force – that is, the level of force necessary to achieve the political objective. Much of the debate in Washington has turned on the circumstances in which the United States should act unilaterally, and those in which it should contribute to multilateral actions. US Ambassador to the UN, Madeleine Albright, has played an important role in this debate by establishing a series of questions that must be answered before US support is given to a UN force. What is the threat to international security? Can the scope of the mission be defined? Are the necessary resources available? And can a clear end point to the mission be established? These epitomised the Clinton approach, summed up in the May 1994 Presidential Decision Directive (PDD) 25 on US involvement in multinational peacekeeping and in the July 1994 White House statement on national security strategy which laid out the administration's broader views on the use of force.

Body-bags and Public Opinion

The need to rally public support for overseas military deployments is a central feature of the US debate. This clearly reflects the trauma of Vietnam; a quagmire that sucked in more and more young US lives to little purpose. In purely military terms, the ghost of Vietnam was in large part laid to rest by the victory of the US-led coalition in the 1991 Gulf War. The highly professional US military felt pride in its achievement. But the manner in which it overcame the Vietnam trauma held serious problems for the future. Both politicians and the public revelled in the technological prowess displayed by the US forces. Indeed, the whole world marvelled at video images of smart bombs passing through windows and cruise missiles turning street corners. But this was a war without pictures of collateral damage. It was brief. And above all there were few US casualties. It was, at least as far as the US public was concerned, a conflict without blood and spilled guts. And it created false expectations as to how the course of US military operations might proceed in future interventions.

Not surprisingly then, the US has been reluctant to commit its forces to messy and unpredictable operations. And when challenged, discretion has often been the better part of valour. The deaths of US servicemen in Somalia in 1993 only served to elicit a clear US timetable for withdrawal. And in Haiti a few weeks later, a robust demonstration by local thugs induced President Clinton to turn away from the harbour a US Navy amphibious landing ship that was about to land armed US troops.

The Clinton administration has wrestled with the question of public support for military action overseas. PDD 25 sets as one criterion whether 'domestic or Congressional support exists or could be marshalled for such a mission'. The July 1994 National Security Strategy statement asks whether a particular mission might have 'reasonable assurance of support from the American people and their elected representatives'. In practice, with the Republican-controlled Congress eager to assert its interpretation of the public mood, such support seems tenuous at best. Most Americans, in Congress' view, do not want their children to return home in body-bags.

This, of course, is a caricature of US public opinion. Like all good caricatures it contains a measure of the truth, but academic studies of polling data suggest that the full picture is far more complex. Steven Kull of the Center for International and Security Studies at the University of Maryland, writing in the Winter 1995–96 issue of *Foreign Policy*, suggests that the polling data are ambivalent and do not support a simplistic assertion that US public opinion is parochial and isolationist. Providing other countries shoulder their share of the burden, most Americans are willing to make a modest, but significant, contribution to international efforts. Indeed the polls show considerable support for multilateralism despite the disasters of Somalia and Bosnia. Kull also questions the simplistic assumption that body-bags will inevitably lead to calls for withdrawal. Indeed, the first US fatality in Bosnia seemed to pass without undue attention; perhaps a function of the fact that it was caused by an accidental mine explosion, that the mission had a clearly established duration and that up to that time, in purely military terms, it seemed to be a success.

Perhaps the most striking finding from the US opinion polls is that many believe that their country's contribution to international military efforts is dramatically greater than it really is. Instructively, they also believe that the general public is far less supportive of a leading US role in international affairs than their own answers suggest. It is in part this negative self-image of public attitudes that generates the stereotypical outlook of isolationism and parochial concerns. Here, however, perceptions are as important as reality. If political leaders perceive the US public as leaning in this direction, policy will be heavily influenced.

Self-deterred Allies?

Other Western nations have come to reflect this 'self-deterrence' to vary-ing degrees. At first sight neither Paris nor London could be accused of combat reluctance. Both served alongside the United States in the Gulf. Both sent significant contributions to the UN Protection Force (UNPROFOR) in Bosnia. And both are playing a leading role in the more robust NATO-led IFOR. Yet if the European military powers are not self-deterred in quite the same way as the United States, their freedom of action is certainly circumscribed. To an extent, this is a question of resources and diplomatic weight; in part a reflection of Europe's peren-nial muddling through when faced with the need to assume common positions, of which the Bosnian crisis is only the most recent and dramatic example.

There have been no sustained calls in either the UK or France to 'bring the boys home', despite a constant trickle of casualties. This may partly be due to the very different foreign policy-making processes; par-liaments in both countries have only a very restricted role compared to that of the US Congress. But the fact that there has been no single incident resulting in major casualties, together with a wide understanding of the very limited nature of the troops' mandate, has calmed any public un-ease. Indeed, successive opinion polls carried out in the UK by Gallup have shown consistently high levels of public support for the British troops' role in Bosnia. Well over 60% of those questioned believe that the UK should play a prominent part in any international force trying to enforce a settlement in Bosnia, and there is strong support for the involve-ment of the UK and other European countries in the crisis as an alterna-tive to letting the various warring factions fight it out amongst them-selves. Such missions may well be seen by large sections of public opinion as an affirmation of the UK's (and probably of France's) role as a leading player on the world stage.

However, the willingness to deploy soldiers abroad should not be confused with belligerence. Even the UK's much-vaunted participation in the Gulf conflict appears to have been seriously circumscribed. The overall British commander in *Operation Desert Storm*, Sir Peter de la Billière, noted in a January 1996 BBC television documentary that fears about unneces-sary casualties explained his reluctance to see British troops attack Kuwait City, preferring instead a wide armoured sweep to the west. Sir Peter was quoted as saying:

> I didn't think we should be losing a lot of British lives. We were there to support a friendly nation and protect their borders. I wasn't prepared to lead a force with monumental casualties [to] victory at the end of the day.

In the Balkans there was similar reluctance to incur losses when genuinely 'vital' interests were not at stake. Successive UK and French governments stresssed the humanitarian aspect of the mission and resisted more overt military action in Bosnia until the international community was humbled and to some extent galvanised by the fall of the proclaimed 'safe haven' of Srebrenica to the Bosnian Serbs in July 1995. A few weeks earlier, military chiefs and defence ministers had met in Paris to set in train the establishment of a robust rapid reaction component for the UN force. Few commentators may have thought it at the time, but the London Conference, also held in July 1995, represented a significant toughening of the international community's line; a decision which in turn made possible the NATO air offensive against the Bosnian Serbs, *Operation Deliberate Force*, that began on 30 August.

This, too, was limited force deployed in a limited way, however deliberately, and it was essentially a political demonstration. Many sorties returned with their bomb racks full for fear of causing civilian casualties. Nonetheless, in contrast to the previous policy, it represented a sea-change in the international community's handling of the Bosnian Serbs.

The two countries which in terms of economic strength have the strongest credentials to act on the world stage – Germany and Japan – have suffered from an acute form of combat reluctance. A better term might be combat-disqualification, a function of their expansionist past and their defeat at the end of the Second World War. In the contemporary world, however, the willingness to deploy force overseas, whatever the constraints on subsequent action, is increasingly seen as part of the ticket to international influence. Both Germany and Japan aspire to permanent seats on the UN Security Council; something that would be unthinkable, at least in the longer term, if they were unable to meet some of the UN's military, as well as financial demands.

The Gulf War highlighted the significantly changed expectations of Germany held by its Western allies after the Cold War. The war in the Gulf, and Bonn's essentially financial contribution to the allied coalition, underlined Germany's limited strategic horizons and the inadequacy of Bonn's old policies. The German Constitution had long been interpreted as preventing the despatch of military force beyond the NATO area. The focus of government activity was to try to build a new consensus that would allow German troops to participate in future multilateral operations under UN auspices.

Germany has moved relatively swiftly to adopt greater international military responsibility. In 1993, a contingent of German soldiers was sent to Somalia. And in a critical judgement on 12 July 1994 the Federal Constitutional Court in Karlsruhe clarified the constitutional basis for the deployment of German forces abroad. German troops could be used for humanitarian missions, broadly defined, providing each deployment received parliamentary approval.

Later the same month, a special session of the German parliament approved Germany's existing participation in the NATO/WEU embargo operation in the Adriatic and to the involvement of German airmen in joint NATO crews involved in AWACS monitoring of Bosnian air-space. In 1995, Germany went a step further in approving the participation of Luftwaffe electronic-warfare *Tornado* aircraft to support the newly created Anglo-French rapid-reaction force in Bosnia. Germany's growing willingness to become involved overseas has prompted practical as well as constitutional preparations. Germany's new defence White Paper published in April 1994 proposed the establishment of a new rapid-reaction force manned by professional, volunteer soldiers. This force would be specifically tailored for international crisis management and peacekeeping missions.

While a new consensus has by no means emerged, the *Bundestag* vote at the end of 1995 that agreed to send 4,000 troops to the former Yugoslavia was instructive. The measure won the support of many of the opposition Social Democrats as well as some of the Greens. Nonetheless, the legacy of the Second World War still places constraints on German action; its forces, largely engineering and logistical units, have thus been based in Croatia.

History and constitutional considerations arguably place even greater constraints on the other world economic power, Japan. But it too has sought to escape from some of the shackles of its past; a booklet published by the Ministry of Foreign Affairs in Tokyo in August 1995 on the United Nations and Japan was sub-titled, tellingly, 'in quest of a new role'. One sign of this new role came in February 1996, when 45 Japanese troops arrived in Damascus to replace Canadian soldiers in UNDOF – the UN Disengagement Observer Force; a small indication of Japan's growing willingness to take on the burden of at least some UN military duties.

Japan's position during the Gulf conflict, like that of Germany, came under intense international scrutiny; there was much implicit, and some explicit, criticism that it was attempting merely to bank-roll its way through the international crisis. Prior to 1992, Japan strictly adhered to the principles of its 1946 Constitution which prevented the projection of military forces abroad for any purpose. Japan's participation in UN operations was restricted to election observers and political officials. The international uproar over its role in the Gulf War pushed the Japanese parliament in 1992 to enact the International Peace Cooperation Law, which allowed members of the Self-Defense Forces to serve overseas under strict conditions. These effectively limit Japanese troops to non-combat situations, monitoring cease-fires or providing humanitarian or logistical assistance.

The first practical deployment of Japanese troops came the same year when a military construction unit of some 600 men was sent to Cambodia. The following year, Japanese military traffic police were sent to Mozambique. In early 1996, a contingent was deployed to the Golan Heights to take part in UNDOF.

If both Germany and Japan have shifted their positions towards peacekeeping in ways that were unthinkable less than a decade ago, their current stance hardly makes them the most likely volunteers to project force abroad. Both countries' freedom of movement is inevitably restricted as much by historical as by domestic considerations. If both the UK and France are reluctant to see their forces committed to combat, except in the most localised scenario, Germany and Japan do not even approach this relatively advanced threshold of reluctance. Japan, for example, declined to send troops to help insulate the Former Yugoslav Republic of Macedonia from the Bosnian conflict.

An Unresolved Dilemna

The idea that a growing and pervasive combat reluctance is afflicting not only Washington, but also to varying degrees its principal allies who look to it for leadership, provides a useful label to cover certain developments over the past half decade. In this reading of events, the 1991 Gulf conflict provides a momentary high-point, after which it is all downhill. But this is perhaps too simple. The essential problem in the post-Cold War world is the absence of any existential threat; hence the difficulty in defining where vital interests lie. And if force is to be committed for lesser 'national interests' then it is much harder to determine exactly when, where and under what conditions this should be done.

The problem is greatest for the one remaining superpower which could claim that virtually every international crisis or conflict in some way strikes at its national interests. The expectations and demands of a sometimes fickle public are of little assistance here. The pervasiveness of the media, and especially of international television coverage and instant commentary, hardly helps. Television, after all, is very good at covering the death of one US soldier. But it is far less good at explaining the many civilian deaths that may have been prevented in central Bosnia by UN-PROFOR's deployment. Political leaders cannot simply court popularity. George Shultz noted wisely in January 1986 that the essence of statesmanship was 'to see a danger when it is not self-evident; to educate our people to the stakes involved; then to fashion a sensible response and to rally support'.

The debate on the use of force in international affairs is set to continue. For those countries that are actually being asked to put their troops 'in harm's way' the dilemmas are all to real, the sign-posts and markers all too indistinct. The Gulf War established an unhelpful paradigm; a bench-mark against which all but the despatch of several heavy divisions may appear as combat reluctance. But the problems remain. NATO has clearly learnt from some of the mistakes of the UN operation in Bosnia. IFOR was planned as a robust force, well-equipped and fully capable of tackling any challenges to its presence. Has its deployment signalled an end to the phenomenon of combat reluctance? Probably not. It is, after all, only supposed to be there for one year.

Arms Control Faces An Uncertain Future

There was intense activity in the area of arms control in 1995. For those in the field, this came as no surprise. For those outside, who may have been lulled into thinking that the need for arms control had greatly diminished since the end of the Cold War, the extension of the Nuclear Non-Proliferation Treaty (NPT), France's resumption of nuclear testing and China's continued nuclear testing, along with concerns over Russian compliance with the CFE Treaty, must have seemed like a blast from the past.

Yet the need to maintain arms-control efforts is great, while prospects for positive advances are uncertain. Even maintaining the advances made in the past is being called into question. Sustaining non-proliferation efforts, even with the passage of the NPT, and concluding a comprehensive nuclear test ban treaty (CTBT), will require constant effort by the nuclear powers – and some appear to be losing their will to make such efforts. Adapting Cold War treaties – such as the Strategic Arms Reduction Talks (START) II Treaty, the Anti-Ballistic Missile (ABM) Treaty and the CFE Treaty – to a very different international environment is another challenge.

Multilateralism is much more important now than it was during the Cold War. Although there is a new convergence of policy among the nuclear-weapon states, China often stands aloof from the consensus. The power of the Non-Aligned Movement (NAM) in the disarmament sphere has also waned. This has much to do with the lack of leadership within the NAM: Mexico had to mute its opposition to US policy for economic reasons, Indonesia lost moral leadership, and India's intransigence on NPT membership – whilst simultaneously developing a nuclear-weapon programme of its own – has prevented it from leading the NAM in the nuclear field.

Bilateral agreements were still a focus of attention in 1995. Not only was the stage set for the ratification of START II by the US and, perhaps, eventually by Russia, but START I came into force in late 1994, with the first inspections in early 1995. The joint US–Russia Agreement for Cooperation on the exchange of classified and sensitive information on nuclear material and facilities was negotiated in April 1995 in Moscow, but its implementation has been fitful. The data exchange agreement being negotiated under the Safeguards, Transparency and Irreversibility talks stalled in November and little progress is expected to be made until after Russia's June 1996 presidential elections.

NPT Extension April–May 1995

The arms-control year was dominated by the effort to extend the NPT. The Treaty was signed in 1968 and came into force in 1970. It contains a clause requiring a decision on its lifetime to be taken at the end of 25 years. There were three available options: indefinite extension; extension for a fixed period; or extension consisting of a series of fixed periods. By January 1995, at the final preparatory meeting in New York, the positions were beginning to harden, but it was still unclear how much support there was for any particular option.

By the beginning of the NPT Review and Extension Conference (NPTREC) in New York on 17 April, the picture had shifted. It became apparent that a majority of the then 179 State Parties to the Treaty favoured indefinite extension. A number of key states (notably Egypt, Indonesia, Venezuela and Nigeria), however, were adamantly opposed to indefinite extension, although they could not agree among themselves on another option. Some preferred a single, fixed extension of between five and 25 years. There were those that preferred rolling extensions of five years, ten years or 25 years. There were also two variations on this option: either that, on expiry, the period would be renewed automatically unless blocked by a Treaty member; or that the renewal of the rolling period would be decided by a vote of the State Parties at each of the expiry dates.

The main reason advanced by those in favour of short-term extensions was concern about the lack of progress towards nuclear disarmament. Under Article VI of the NPT, all states were committed to 'pursue negotiations in good faith on effective measures relating to cessation of the nuclear arms race at an early date and to nuclear disarmament, and on a treaty on general and complete disarmament under strict and effective international control'. Even though the US and Russia could point to the Intermediate-range Nuclear Forces (INF) Treaty and to START I and II, and all states could point to the CTBT negotiations and agreement on negotiations to ban military fissile material, some argued that there were more nuclear weapons in 1995 than there had been in 1970, and claimed that the three lesser nuclear powers (the UK, France and China) and the three 'undeclared' or 'threshold' states (Israel, India and Pakistan) had made little contribution to ending the nuclear arms race. In addition,

Egypt raised the stakes over Israel's non-participation in the Treaty and linked its extension, of any duration, to Israel joining the NPT.

Even within those states that favoured indefinite extension there were differences of opinion. These hinged on the policy of indefinite and unconditional extension promoted by the US, France and the UK. Although virtually all Western states, Russia and East European states, plus many other allies of the US, UK and France were strongly in favour of extending the NPT indefinitely (for reasons of regime stability and long-term concerns over proliferation), many of these states were uncomfortable with the idea of imposing no conditions on the nuclear-weapon states, particularly with reference to Article VI of the Treaty.

It was generally felt that if the extension decision were pushed to a vote, there would be a small majority in favour, but there would also be a sizeable minority that felt disenfranchised within the Treaty. Some also believed that the nuclear-weapon states should not be allowed to achieve an unconditional indefinite extension of the Treaty; they should instead be required to make some concessions to the concerns of other states on nuclear disarmament.

Four major factors led to the final decision on NPT extension. Of great importance was the fact that the Conference Chairman, Jayantha Dhanapala, Sri Lanka's Ambassador to the US, not only understood the issues clearly, but also supported the NPT and nuclear disarmament. He represented a state with a solid reputation within the NPT and he was a tough negotiator who, from the outset, set his sights on compromise. He was respected by all sides.

Second, the NAM was not united. Different factions within the movement put forward different proposals, whereas the Western and Eastern Groups had only one position: indefinite extension. In the third week of the NPT conference, the non-aligned states held a conference in Bandung, Indonesia, but even there they could not hammer out a unified position.

Third, the Arab states negotiated a resolution that was then proposed by the three depository states (the UK, the US and Russia) calling on all states in the Middle East to establish a zone free of weapons of mass destruction.

Fourth, South Africa, newly emerged onto the world stage and fortified by its credentials as a state which had renounced its nuclear weapons, played a key role. It joined forces with Canada to promote an effective compromise that would allow an indefinite extension with some conditions. Whilst Canada collected signatures on a motion in favour of indefinite extension (so that a large majority in favour could be clearly demonstrated), South Africa worked on obtaining support for a document that became the 'Principles and Objectives' package. Of particular significance are the references within this package to nuclear disarmament, safeguards and peaceful uses of nuclear energy.

• *Nuclear Disarmament:* The document refers to easing international tension, strengthening trust between states and fulfilling their undertakings with regard to nuclear disarmament. Specific objectives include concluding an internationally and effectively verifiable CTBT no later than 1996; early conclusion of a ban on the production of fissile materials for nuclear weapons; and systematic and progressive efforts to reduce nuclear weapons globally, with the ultimate goal of total elimination.

• *Safeguards:* The document reaffirms the role of the International Atomic Energy Agency (IAEA) to verify compliance with the NPT. It specifically states that IAEA safeguards should be regularly assessed and evaluated, although it does not say by whom. The document also states that decisions adopted by the IAEA Board of Governors, aimed at further strengthening the effectiveness of its safeguards, should be supported and implemented, and that the Agency's capability to detect undeclared nuclear activities should be increased. Significantly, the document recommends that nuclear fissile material, when transferred from military use to peaceful nuclear activities, be placed under safeguards in the framework of voluntary agreements in the nuclear-weapon states, and that safeguards be universally applied once nuclear weapons have been completely eliminated.

• *Peaceful Uses:* The document reiterates the inalienable right of States Parties to develop nuclear energy for peaceful purposes. Its stress on giving preferential treatment to non-nuclear-weapon states was tempered by advocating transparency in nuclear-related export controls within the framework of dialogue and cooperation among all interested States Parties. The package referred specifically to the standards of accounting, physical protection and transport of nuclear materials, and to the adequate resourcing of the IAEA to ensure it meets its responsibilities.

As a result of this package deal, the NPT was extended indefinitely (if not strictly unconditionally) on 11 May 1995 without a vote, although significantly not by consensus, thereby bypassing the divisive and, in the long term, dangerous likelihood of a split vote with a small majority.

There are a number of important measures within the document. The most important are the signing of a CTBT by 1996, progress towards eliminating nuclear weapons, and strengthening the IAEA safeguards programme. All these issues are to be discussed within an enhanced review process, whose first meeting will take place in 1997 and thereafter every year until the next Treaty Review meeting in 2000. If, by the 1997 meeting, a CTBT is not in place, or if the IAEA still cannot implement all of its enhanced programme, or if there has been no movement towards negotiations of total nuclear disarmament, the NPT will come under heavy attack from those who set these goals as the condition for their agreement.

Negotiations in Geneva

Nuclear Test Ban

The Conference on Disarmament (CD) has been negotiating in Geneva for a CTBT since January 1994. Agreement on a number of key issues was reached in 1995, but some major ones had still to be settled. Much remains to be done in resolving disagreements if the CTBT is to be ready for signature by the UN General Assembly in November 1996.

Ideally, agreement on a CTBT would have been reached before the April–May 1995 NPT Review and Extension Conference, but by March 1995 it was clear that this was not feasible. Nevertheless, in an attempt to show good faith, the UK and France in early April dropped their insistence that safety tests be allowed within the CTBT. In addition, at the NPTREC the nuclear-weapon states pledged 'utmost restraint' in nuclear testing prior to signing the CTBT in 1996. The hollowness of this pledge was revealed when China conducted its 42nd nuclear-weapon test on 15 May, less than 48 hours after the Conference ended. On 13 June, newly elected French President Jacques Chirac created further consternation when he announced the resumption of France's nuclear testing programme. After a moratorium of over three and a half years, France was to conduct a series of eight tests (in the end only six) in the South Pacific. The international reaction to China's nuclear tests in May and August, and to France's tests from September to January, was severe. This was particularly true in the Pacific, with Australia, New Zealand and Japan leading the protests. The French government was also taken aback by the strong reaction to the tests within France – there were demonstrations on the streets in Paris – and in some other European Union states, not least Germany.

At the same time in Geneva, the nuclear-weapon states were discussing the issue of treaty scope. The US wanted to carry out very small nuclear explosions (hydro-nuclear tests), while France, the UK and Russia were keen to extend such an exception to carry out much larger explosive yields – up to a few hundred tons equivalent of TNT. These developments caused international uproar. Many of the non-nuclear NPT member-states felt betrayed. Those who had opposed indefinite extension felt vindicated in their approach; the pressure from states with influence and from key non-governmental organisations focused minds. On 10 August, President Chirac announced that France would now propose a 'truly comprehensive test ban', and the following day US President Bill Clinton announced that the US would seek a 'zero-yield CTBT'. In September, the UK joined the zero-option trend and, at a meeting in October in New York between President Clinton and his Russian counterpart, Boris Yeltsin, it was announced that Russia, too, was in favour. Russia has yet to state this position formally in Geneva, however.

China's position on scope is ambiguous. China had proposed a 'no release' test ban. Yet it is insisting on allowing so-called 'peaceful nuclear explosions', a position unsupported by any other negotiating state. In addition, China is against including data from national technical means (including military satellites) in the international monitoring system. China has also not indicated whether it intends to stop its nuclear testing programme on signing the CTBT or when it enters into force.

There is thus still much confusion over what the permitted activities will be under a CTBT, and India in particular is exploiting this confusion. In December 1995, the US government claimed to have evidence from intelligence satellites that India was preparing a second nuclear test (the first was in 1974) at Pokharan in the Rajathstan desert. While India has denied this vehemently, it has insisted on linking the CTBT with a commitment to a time-bound framework for total nuclear disarmament. This approach will probably fail. India has garnered little support, and some believe that it is simply trying to escape from the CTBT, ironically a treaty first proposed by Nehru in 1954. Nevertheless, even if India does not scupper the CTBT, it may delay it. A CTBT may still be achieved in 1996, but it will require forcing China and India into line or leaving them out altogether. Neither option will guarantee a swift or easy passage to signature or ratification.

Ban on Fissile-Material Production

There was considerable excitement when a mandate was agreed in Geneva just prior to the NPTREC to negotiate a ban on fissile-material production for military purposes. It was premature, however, for very little has since been achieved. Severe disagreement over the wording in the mandate arose, with some countries, particularly Pakistan, Egypt, Iran and Algeria, insisting that stockpiles of fissile material be explicitly included in the mandate. Other states, such as the UK, the US, France, Russia and India, were adamant that stockpiles should not be mentioned. On 23 March 1995, a compromise was agreed whereby the mandate would not specifically include stockpiles, but nor would they be barred from negotiations either. That is as far as the negotiations went, however. The committee to negotiate a fissile material ban has yet to be convened and the situation is in stalemate. It is unlikely that there will be action on a ban until after the CTBT has been negotiated.

Nuclear-Weapon-Free Zones

There was a generally favourable trend in 1995 in agreeing and signing nuclear-weapon-free zones (NWFZ). Many argue that such zones are merely symbolic, but that underestimates the importance of symbolism. Beyond the symbolism there is a harder reality. All the NWFZ treaties now in place have some form of verification regime and, in the near future, all will have protocols signed by at least four of the five nuclear-weapon states. Article VII of the NPT directly encourages the establishment of the NWFZ, and this was reinforced at the 1995 Review Conference. In 1995, the 1974 Treaty of Tlatelolco was ratified by Guyana, St Kitts and Nevis, and St Lucia. In March, Cuba also signed the Treaty, hence all states within the zone of application have now signed. All five nuclear-weapon states have also ratified the relevant protocols.

France's use of its nuclear test site at Mururoa Atoll in the South Pacific rewakened the strong anti-nuclear feeling exemplified in the 1985 Treaty of Rarotonga which proclaimed the South Pacific a nuclear-free zone. In reaction to the uproar that ensued, France, the UK and the US jointly announced that, following the end of French tests in the region, they would ratify the relevant protocols of the Treaty, and so they did in late March 1996. This welcome move came after the three allies had held out against the Treaty for years, not only because of France's nuclear testing programme, but also because of the Australia–New Zealand–US (ANZUS) alliance and the deep-seated fears of 'nuclear allergy' in the region following New Zealand's ban on visits from nuclear-armed or nuclear-propelled vessels.

The African Nuclear-Weapon-Free Zone was submitted to the 50th UN General Assembly for approval after it was finalised in June 1995 in Pelindaba, South Africa. It is thus generally known as the Pelindaba Treaty even though it will be signed in Cairo in 1996. The nuclear-weapon states have had some difficulty in agreeing to sign the protocols because territory in the zone of application is used for military basing (Diego Garcia in particular). Despite this, the UK and the US are expected to sign at least two of the protocols.

The movement has now been extended to Asia. On 15 December 1995 in Bangkok, the Association of South-East Asian Nations (ASEAN) states agreed and signed the South-East Asia Nuclear-Weapon-Free Zone (SEANWFZ). The nuclear-weapon states have problems with legal interpretations of the protocols, however, and these are delaying their ratification. China is concerned about the language regarding the South China Sea, part of which is included in the zone of application, and the US is concerned over the effect it will have on the passage of nuclear-powered or armed vessels.

START I and II

The 1991 bilateral START I agreement finally entered into force at the end of 1994. The instruments of ratification from Russia, Ukraine, Belarus, Kazakhstan and the US were exchanged in Budapest on 5 December 1994. In January 1995, the US and the other four signatories supplied data to each other as required by the Treaty. The numbers involved indicated that both sides had been reducing their weapons during the long waiting period before START entered into force.

At the beginning of March 1995, the START baseline inspection period began. For the following 16 weeks US teams conducted 74 inspections, including baseline, close-out and elimination inspections. In addition, there had been continuous monitoring at two production facilities since January 1995.

Inspection teams representing the other four states carried out 37 inspections within the US. Only one of these inspections was not carried out by an all-Russian inspection team, but by a mixed Ukrainian–Russian team, led by Ukrainians. By the end of 1995, following the baseline inspection period, the US had carried out 37 inspections at former Soviet sites and the four states had carried out 24 inspections at US sites. Russia had not exercised its right to monitor continuously the MX final assembly plant in Promonotory, Utah.

The 1995 inspection process was hailed as a great success, as was the rate at which weapons were dismantled. By the end of 1995, both sides were well below the strategic nuclear-delivery vehicle level required by 1999; Kazakhstan had become nuclear-weapon free; after a temporary hitch, Belarus was back on track in transferring its nuclear weapons to Russia before the end of 1996; and Ukraine reported that over 40% of the nuclear weapons on its territory had been shipped to Russia; as long as Russia continued to deliver fuel rods for Ukraine's nuclear power plants, all would be transferred before the end of 1996.

The only difficult issue within the START I implementation process was that of using missile stages for space-launch vehicles. Such a conversion process is allowed, in a limited fashion, by the Treaty, but notice must be given. In order to reconcile these difficulties the Joint Compliance and Inspection Commission issued statement number 21 at the end of September firming up the agreed terms under which conversion of rocket stages from missile to space launchers would be acceptable.

1995 was a more difficult year for START II. This Treaty incorporates two phases of elimination, the first running concurrently with START I and the second due to end on 1 January 2003. But it has yet to enter into force, although it was signed at the beginning of 1993. One hurdle was surmounted when the US Congress ratified the Treaty on 26 January 1996, but its passage through the Russian Duma looks increasingly problematic. Not only do the nationalist tendencies in Russia make arms-control agreements with the US a popular stick with which to beat the democratic reformers, but concerns over NATO expansion, the future of the ABM Treaty, and the costs of dismantling nuclear weapons also play a large part in bolstering opposition.

In any event, the Duma has had very little time to consider START II properly. President Yeltsin did not forward it for consideration until the end of June 1995, and the administration delayed reporting to the Duma on the costs and design of Russian force restructuring under START II. Perhaps the main concern for Russia is the differential in US and Russian forces when START II limits are reached. There is a strong sense in the Duma that the Yeltsin government negotiated the Treaty badly. There are concerns that after START II is implemented the US will still have the capacity swiftly to upload *Minuteman* IIIs, *Trident* IIs and to convert heavy bombers, while Russia will be unable to upload its systems quickly. Against this background, US resistance to advance proposals for START III raises further doubts in the Duma. New negotiations could address the perceived shortcomings of START II and allow the pro-arms-control forces in Russia to demonstrate the commitment of the US to further and, in Russian eyes, more equitable cuts in nuclear arsenals.

In addition, the Duma has very real concerns, shared by the international arms-control community, over the ABM Treaty and theatre missile defences. Russia's new Foreign Minister, Yevgeni Primakov, has taken an increasingly tough line on the proposed agreed statement to clarify the ABM Treaty, which will describe the difference between theatre and strategic missile defences. Nevertheless, negotiations on the technical aspects of demarcation are continuing, and the two sides have managed to make some progress in narrowing the gap between them.

Conventional Forces in Europe Treaty

The CFE Treaty was dogged in 1995 by problems over Russia's requests for increased holdings in its flank limits and to its concerns over NATO expansion. Russia had not achieved compliance with the Treaty by the end of the reduction phase in November 1995 because equipment levels in the North Caucasus were above permitted levels. The issue remains unresolved, although some common ground was found after much negotiation.

In July 1995, at a meeting in Brussels with NATO countries, Russia proposed an exclusion zone in the North Caucasus that would not be subject to CFE Treaty limits. This would not be the first exclusion zone; one already took into account Turkey's specific security concerns. Later that month in Vienna, Russia announced its proposed force levels in the North Caucasus and in the Russian flank zone, all of which exceed the allowed CFE levels. Russia's argument was that the CFE Treaty allows for flexibility in the temporary deployment of excess equipment and that the situation in the Caucasus meets this criterion. In addition, Russia proposed that the storage rules be relaxed, that others within the flank zone yield some or all of their quotas to Russia and that the flank zone be redefined.

NATO responded to the Russian proposals in Vienna in September 1995. It agreed to redraw the boundaries of the flank zone if Russia would accept more inspections and report more information. US–Russian bilateral discussions in New York in October and discussion in Vienna further refined the proposals, but because of Turkish unease the matter remained to be resolved at the May 1996 CFE Review Conference.

Inhumane Weapons Convention

The 1981 Convention on Certain Conventional Weapons (or Inhumane Weapons Convention) came into force on 2 October 1983, together with three protocols on non-detectable fragments, mines, booby-traps and other devices and incendiary weapons. Article 8 of the Convention allows for reviews and amendments. France submitted an official request for a review conference in December 1993 which took place between 25 September and 13 October 1995. On the agenda were anti-personnel mines and blinding weapons. These two issues had been pushed to the top of the list by medical, development and human-rights organisations (such as the International Committee of the Red Cross, Oxfam and Human Rights Watch) which then enlisted the assistance of specialised arms-control non-governmental organisations (NGOs).

The Conference had mixed success. On the positive side, it adopted a protocol on blinding laser weapons which bans their deliberate use and transfer, but not their production. The protocol, however, allows the use of such weapons to target optical systems; any blinding as a form of collateral damage is not covered by the protocol's prohibition.

There was no agreement on the more difficult issue of the production, transfer and use of land-mines. There was agreement on some aspects, such as the need to extend the convention's land-mine restrictions to internal as well as international conflicts, and a prohibition on land-mines that explode when an electromagnetic detector used to clear them comes near. There was much disagreement, however, on the treatment of self-destructing or self-neutralising mines, and the adaptation, and timing of adaptation, of mines to achieve the humanitarian objectives of the Review Conference. Most of the key NGOs and agencies involved in the debate are promoting a total ban on land-mine production, transfer and use. The Review Conference was scheduled to continue in April–May 1996.

The Wassenaar Accord

As the successor to the Coordinating Committee for Multilateral Export Controls (COCOM), the Wassenaar Accord has been a long time in the making. COCOM was created in 1949 to prevent the transfer of militarily sensitive technology to the communist bloc. The Wassenaar Accord, signed on 19 December 1995 in Vienna, includes many of the countries targeted by COCOM (such as Russia, Hungary, Poland, the Czech Republic and Slovakia). There are currently 28 member-states and the agreement is, as expected, much weaker than COCOM. Compliance with the Accord will be voluntary, export decisions will be taken by each member-state individually, and arms transfers have to be disclosed only after the fact. Only Iran, Iraq, Libya and North Korea are currently on the 'banned' list for the transfer or sale of arms. The weaknesses in the Accord reflect the difficulties of reaching agreement in today's international environment; in the absence of a clear enemy, the need for compromise has moved to the fore.

Biological and Toxic Weapons Convention

Rapid advances in the bio-technology industry have made biological weapons a much more practical proposition than in the past. These developments have emphasised the importance of the task given to a working group of parties to the Biological and Toxic Weapons Convention (BTWC) which struggled throughout 1995 to draft a verification protocol. The draft protocol is intended to be ready for the next BTWC Treaty Review Conference in December 1996. Startling revelations in August 1995 of the scope and nature of Iraq's offensive biological weapons programme highlighted the acute need for a strong verification regime. The defection of Saddam's son-in-law, General Hussein Kamel Hasan, spurred Iraq into releasing a mass of new detail on its weapons programmes and heightened concerns about the threat associated with these weapons.

At present the BTWC has no verification provisions. The working group has made some progress in identifying possible measures, but the intrusive verification needed to make it effective poses considerable problems for a global regime. Coverage of a wide range of legitimate civil and government facilities would be needed and protection of proprietary rights, public safety and national security set limits on what is politically possible. Even the most intrusive inspection regime ever implemented in a single country – by the UN Special Commission in Iraq over four years – was able to discover only shreds of evidence about its weapons programmes. Any proposals for a verification regime for the BTWC will have to be carefully analysed to assess their true effectiveness. A political 'quick fix' arrived at for the sake of agreement could damage security rather than enhance it.

The Chemical Weapons Convention

Unfortunately, the 1992 Chemical Weapons Convention has yet to come into force. During 1995, 28 states ratified the Treaty bringing the total to 47 by the end of the year. The UK completed its ratification passage through parliament and should deposit the instruments of ratification by mid-1996, but the US and Russia made little headway during the year. The Treaty requires 65 states to ratify before it can enter into force and states which have not completed the process before entry into force (six months after the first 65 deposits) lose a number of advantages. Specifically, they will be outside the decision-making structures of the Treaty and their nationals will not be employable as inspectors or within the Organisation for the Prohibition of Chemical Weapons. If the US and Russia are not among the first 65 states to ratify, the Convention will be in serious trouble.

Fulfilling Promises

1995 was a critical year for nuclear arms control. In order for the NPT to be permanently extended, the nuclear-weapon states agreed to a more vigorous review process and a strong commitment to a CTBT and a ban on the production of fissile material for military purposes. The next few years will prove the litmus test of these commitments. If the CTBT is not negotiated in time for the first Preparatory Committee of the enhanced review process of the NPT in 1997, there could be a hard road ahead for non-proliferation efforts. In addition to the CTBT, it is vital that the nuclear-weapon states begin negotiations on nuclear disarmament. Currently, most countries are calling for these to be held within the Conference on Disarmament in Geneva, but this may not be the most effective forum. Instead, the five nuclear-weapon states, along with the three undeclared states, could begin negotiations at a separate conference to agree a series of confidence-building and transparency measures in preparation for eventual multilateral disarmament. Whatever they decide, the nuclear-weapon states are being forcefully reminded of their commitment to eliminate nuclear weapons and that procrastination will only serve those states that intend to proliferate.

The process will not be easy. There are many uncertainties, not least the direction that Russia will choose, and its response to the question of NATO enlargement. Russia's approach to the CFE Treaty and the START II ratification debate demonstrated how treaties negotiated in Cold War structures have to be adapted and amended to fit contemporary needs.

Despite doubts about future structures, those that now exist, partly through their verification regimes, are providing all parties with information and confidence on which to build future, realistic defence policies. It is essential to ensure these are adequately funded and designed in such a way as to be cost effective. Existing regimes, such as the IAEA safeguard system, and possible new ones, such as those being considered for the CTBT and the BTWC, should be vigorously analysed from this point of view.

The Americas

The volatile nature of US political opinion was underscored during 1995 and the early months of 1996 as the fortunes of the Republican Party sunk steadily. Having won control of both houses of Congress in 1994, the leadership's attempts to force radical ideological legislation into law backfired. President Bill Clinton used his power of veto effectively to convince the public that he was protecting their rights. His standing in the polls rose, and he faced an election year in far better shape than most had thought possible at the start of 1995.

Unable to initiate domestic programmes because of the Republicans' grip on the legislature, Clinton became a foreign-policy activist. He provided a necessary balance through his use of US power in the Bosnian crisis and through the threat that US force might be used in the Taiwan Strait. Both these actions were at least temporarily successful. His support for peace in Northern Ireland and the Middle East, however, was upset by renewed fundamentalist violence. If further progress is to be made, it will be important to keep the US involved in these crises.

The US was also caught up in the security tensions in Latin America. New strains developed in its already difficult relations with Cuba; its role in the anti-drugs war adversely affected its relations with a number of Latin American countries; and the turmoil in Mexico's economic and political life made for trouble between the US administration and Congress. By side-stepping Congress, however, Clinton provided sufficient credit and financial support to help Mexico surmount its extremely difficult economic situation. Full stability in Mexico and elsewhere in Latin America is far from assured, but at least most of these countries lost no ground in 1995.

The US: Leading Abroad, Troubled At Home

During 1995, US politics became ever more perverse. President Bill Clinton, who had come to power in 1993 on a platform of radical change in domestic policy, but as a supposed foreign-policy novice, found himself practically excluded from the domestic agenda and focusing heavily – with considerable success – on foreign policy. Back at home, the first Republican-dominated Congress in 40 years was overwhelmingly elected in late 1994 amid talk of a 'revolution', but by the end of 1995 it had passed very little legislation and was just as unpopular as the Democratic Congress it had replaced.

When the Republican 'revolutionaries' became frustrated with their inability to achieve their primary goal – balancing the budget by cutting spending – they attempted to force the President to accept their demands by refusing to extend the temporary budget authority under which the government had been operating in the new fiscal year. As a result, much of the federal government was shut down. Clinton, however, refused to yield, and 1995 ended with nearly 800,000 federal workers at home and unpaid, national parks and museums closed, passport and visa applications piling up at home and abroad, and both branches of government increasingly unpopular. Adding to the frustration of the Republicans in Congress, Clinton foisted most of the blame onto them.

The ups and downs and ironies of 1995 might have been amusing had it not been for their implications for thousands of individuals, the respect for the government throughout the country, and the reputation of the United States throughout the world. With 1996 an election year, the domestic political circus in Washington will make US policy unpredictable for at least another year at a time when the world needs leadership that can only come from the United States.

Domestic Politics

President Clinton's domestic political fortunes reached a low point at the end of 1994. His main domestic-policy goal, health-care reform, had failed, and his policies as a whole were repudiated in legislative elections that cast out the prevailing Democratic majority. The Republicans took control of both houses on a platform of cutting taxes, balancing the budget, curbing welfare programmes, implementing Congressional term-limits, and increasing spending on national defence. Their election seemed to mark an ideological turning-point in US politics, a definitive rejection of the government activism that Clinton represented, and an overwhelming acceptance of the opportunistic conservative politics of the new Speaker of the House, Newt Gingrich.

By the end of 1995, however, this reading of events was no longer as convincing. Although the Republican agenda was popular, Republican leaders had difficulty producing a formula that would do all they wanted to at once. Unwilling to forego their planned cuts in taxes on capital gains, on payments to middle-class families with children, and on social-security benefits, House Republicans concentrated their cuts on welfare spending. In March 1995, they proposed bills to cut poverty programmes by $70 billion over five years, cap government support for teenage mothers, and move welfare spending from the federal government to the individual states, with limits on how much they could spend. The more cautious Senate was unwilling to back all these proposals. Nevertheless, in June, Senate leader Bob Dole reached an agreement with Gingrich to cut government spending by nearly $1 trillion while also reducing taxes by $245bn. In theory this would result in a balanced budget by 2002. The Republicans had indeed proposed their balanced budget and their tax cuts, but only by threatening drastic reductions in federal spending on Medicare for the elderly, Medicaid for the poor, education, the environment, and help for impoverished families with dependent children.

The Republican proposals forced Clinton to come up with his own counter-proposal. After much wavering he finally agreed on some tax cuts and a (not very specific) balanced budget of his own. But the Republicans made a mistake in trying to use their control over raising the federal debt ceiling to force the President to accept their version of how and when the budget should be balanced, including tax cuts for the rich and major cuts in entitlements. This allowed Clinton to take a stand as the guarantor of the limited social protections that Americans had, and painted his opponents as heartless ideologues who wanted to take away the rights of the middle class and the meagre comfort given to the poor. Clinton found defensive politics more congenial than presenting radical proposals of his own, and it apparently worked: only 34% of Americans polled in late 1995 believed Congressional budget proposals were 'good for America', while 51% thought Clinton's were. Polls also revealed that 49% of Americans now had confidence in the President compared to only 35% who had confidence in the Republicans, a sharp turnaround from the 46% to 30% advantage the Republicans had enjoyed just a year before.

Rather than signalling the arrival of a new and sharply conservative ideology, the 1994 Congressional elections – as the 1992 elections before them – may instead indicate that the US public is dissatisfied with the traditional remedies of both right and left. It now seems more willing to listen to whatever new ideas are on offer, so long as they are the opposite of the principles prevailing in Washington. The elections of both 1992 and 1994, in fact, brought more new members to Congress than at any time since the Second World War, with 110 freshman Representatives and 12 Senators in 1992 and 86 new Representatives and 11 more Senators in 1994. In addition, frustration with politics (along, no doubt, with fear of losing) drove more Representatives and Senators than at any time since the nineteenth century to announce their retirement from office without even bothering to run in 1996. They included the Democratic heavy-weights Sam Nunn, Paul Simon and Bill Bradley, but also Republicans including Alan Simpson, Nancy Kasselbaum and William Cohen, to whom the budget battles of 1995 were the 'last straw'. Most of these retirees were moderates who could no longer put up with the extremism, ideology and lack of civility that now reigned on Capitol Hill. The House

Republicans failed to pass a Congressional term-limits bill, as promised in their 'Contract with America', but the increasing exasperation of legislators, along with the voters' growing desire to throw out whoever was currently in office, proved to be the bill's functional equivalent.

Race continued to play a prominent role in US domestic politics, and was perhaps even more of an issue in 1995 than usual. The trial of former football hero and movie star O. J. Simpson – televised on some cable channels for 24 hours a day – not only kept the country mesmerised (one think-tank estimated the cost of lost US productivity from Americans staying home from work to watch at $27bn, and another reported that the cost of the trial itself was more than the gross national product of several UN member-states), but also revealed a disconcerting racial divide. Whereas an overwhelming majority of whites were convinced that Simpson was guilty of murder, more than 70% of blacks believed he was innocent, the victim of a racist justice system. Meanwhile, more studies documented the plight of black American males, showing that more than one in three between the ages of 20 and 30 were involved with the justice system, and often in prison.

This helped to create the climate conducive to the success of the so-called 'Million Man March' of 16 October 1995 in Washington, attended by hundreds of thousands of black men determined to take control of their lives. Some had been inspired by, and others came in spite of, the divisive rhetoric of the organiser of the march, Louis Farrakhan. Race also figured prominently in Congressional debates, as 'affirmative action' programmes were branded by Republicans as unfair and failed 'social engineering'. One potential race-related bright spot might have been a presidential election campaign by General Colin Powell, a black man widely respected and admired by people of all races, but in early December 1995 Powell announced that he would not run, partially because of the intrusive nature of the US presidential campaign, but also, perhaps, because of the race issue.

If race was an old American problem, anti-government terrorism was a new, or at least growing, one. The country was shocked by the 19 April 1995 bombing of the Alfred P. Murrah Federal Building in Oklahoma City which killed 169 people, including many children. The bombing, it turned out, was not the act of some foreign terrorists as initially assumed, but was perpetrated by US right-wing extremists manifesting their hatred for federal government. It soon became clear that most southern and mid-western US states had armed militias whose self-appointed mission was to stop the 'federalists' from dominating their lives. Along with other examples of anti-government actions and attacks on federal judges, forestry offices and even the White House itself, the Oklahoma City bombing revealed just how alienated from government some US citizens felt. Shootings at the White House led to the closing of Pennsylvania Avenue to traffic for security reasons which only reinforced, and sadly symbolised, the growing distance between 'Washington' and the rest of the country.

Distracted by these problems, few observers seemed to notice how well the US economy was doing. Growth in gross domestic product (GDP) was 3%, marking the fourth consecutive year of uninterrupted expansion. Unemployment remained under 6% (near the rate of 'full employment'), and inflation stayed low at 2.5%. The budget was, of course, still not balanced, but the deficit was decreasing and, at 2.4% of GDP, stood at a level that most industrialised countries would envy. Corporate America continued to thrive, and the stock market broke new records, rising by more than 33% for the year. The main blight was not, as implied in much Congressional debate, the budget or even the trade deficit, but the continued stagnation of wages among middle- and low-income workers. Even though overall national income continued to rise, median income remained about the same as in the early 1970s. This was a consequence of the growing inequality of income, despite the Clinton administration's increased taxes on the highest earners. Republican proposals to cut government spending, end aid to many welfare recipients, and flatten out the progressive tax system might fulfil their claim of more overall economic growth, but were certain to exacerbate national differences in income. This 'fairness issue', in fact, was one of the main points dividing President and Congress in the budgetary debates.

Taking Risks for Peace

The first two years of President Clinton's foreign policy were marked by hesitation, uncertainty and caution. Retreats from announced positions on Somalia, China, Haiti, the UN and Bosnia seemed to suggest that foreign affairs were not much of a US priority at all. Yet in 1995 the President apparently began to grasp that foreign-policy leadership can actually contribute to political capital rather than decrease it, and the United States began to demonstrate the leadership and commitment that only it could offer. Particularly in the former Yugoslavia, the Middle East, Northern Ireland and Haiti, US assertiveness and willingness to pursue peace actively contributed significantly both to developments on the ground and to preserving the reputation of the United States as a world leader. US leadership, of course, was not the only reason why peace advanced in certain parts of the world, and in some cases the Clinton administration simply took the credit for welcome developments that occurred while it was in office. Still, a strong desire for US mediation in international disputes undeniably existed, and in 1995 the US government filled this void more effectively than in the previous two years.

The main foreign-policy preoccupation for the administration in 1995, as in the previous two years, was the war in Bosnia. But this year the US was more involved and with far greater success. After a 60-day cease-fire was agreed in early October, negotiators from all the warring parties attended peace talks under US aegis at an air base in Dayton, Ohio. On 21 November the Dayton Peace Agreement was initialled, calling for a partition (although the word was not used) of Bosnia and Herzegovina into two separate 'entities', the exchange of territory according to a negotiated map, and the deployment of a massive NATO Peace Implementation Force (IFOR) to ensure compliance.

To some Americans (and to others as well) the Dayton agreements revealed that the best recipe for diplomatic success is for the US to take the lead while others, including the divided and 'pusillanimous' Europeans, leave well alone or at most follow passively. This view overlooks the important role played by US allies such as France in stiffening NATO's resolve and implementing its decisions; the fact that at Dayton the US more or less accepted the European partition plan that Washington had long refused; and, most importantly, the role that developments on the ground in Bosnia played in facilitating the peace. Still, it was US leadership that proved decisive. Only the United States had the credibility, unity and military power – all ably represented in the form of Special Envoy Richard Holbrooke, whose indefatigable cajoling played a key role in its success – to bring the warring parties to the negotiating table. President Clinton's decision, in the face of Congressional and public opposition, to deploy more than 20,000 US troops to Bosnia as part of IFOR was an indispensable element in the package that eventually brought peace, and stood in marked contrast to the US irresolution of 1993–94. The decision to deploy ground troops – taken not only to promote Bosnian peace, but also to show NATO allies that the US was still a reliable leader (a point made explicitly by administration officials and the key selling point to the Republican Congress) – reassured its allies that the US was not turning its back on the rest of the world.

The allies might have been forgiven for fearing that the US was indeed turning away. Not only had Clinton shown an erratic interest in foreign affairs, but the Congress that was elected in November 1994 came to power determined to narrow the definition of US national interests and, more than anything, to reduce the costs of the country's role in the world. It was not so much that the new legislators were 'isolationist' in that they carefully considered, and then rejected, involvement in international affairs. It was rather that they did not think much about the rest of the world at all, except as something that cost US taxpayers money and should therefore be viewed with suspicion. When, in January 1995, the administration proposed a plan to guarantee $40bn of Mexican debt, for example, Congress gave its reluctant support, but not one of the 78 freshmen Republicans could be persuaded to join.

Congressional committees proposed legislation to cut spending on foreign aid and diplomatic activities drastically; to abolish completely four government bodies dealing with international affairs (the Commerce Department, the Agency for International Development, the United States Information Agency, and the Arms Control and Disarmament Agency); to close numerous consulates abroad; and to prohibit US forces from serving under United Nations command. The administration successfully blocked much of this programme (although not the foreign-aid spending cuts or the closing of consulates), and ironically found its own foreign policy largely defined and even boosted by its opposition to Congressional isolationism. Whereas for two years Clinton had been criticised by the Republicans for weakness, lack of engagement and leadership abroad, the administration was now promoting and demonstrating US leadership, most clearly in Bosnia. Congress seemed out of touch.

In the Middle East, too, US leadership was clear. Washington remained the most desirable interlocutor of all the parties to peace negotiations, and Secretary of State Warren Christopher's repeated visits to the region helped to keep the peace process on track. The key to achieving a comprehensive Middle East peace – a deal between Syria and Israel – was significantly advanced when the parties agreed in November 1995 to hold talks under US auspices at the secluded location of Wye Plantation on Maryland's eastern shore. Washington also remained the primary outside actor in the Persian Gulf. The US insisted successfully on maintaining UN sanctions against Iraq and put in place an elaborate system of prepositioned weapons-stock and forces – including the newly created Fifth Fleet based in Bahrain – in case the Baghdad regime should ever be tempted to threaten its neighbours again. US policy towards Iran was less successful: on 2 May the Clinton administration cut off all trade with Iran and called for a total international boycott, but the allies refused to agree and Russia proceeded with a nuclear power reactor deal (allegedly stripped of military components) despite vigorous protests by Washington. Although the 'dual containment' policy offered little hope of transforming the Iraqi or Iranian regimes, it at least helped to isolate them and prevent them interfering in the Arab–Israeli peace process or bothering their neighbours.

Clinton's newly discovered interest in foreign affairs was also evident in Europe, especially during his triumphant December 1995 trip to the United Kingdom. Leaving the squabbling budget negotiators behind in Washington, Clinton looked more presidential than ever as he spoke to a packed House of Commons in London and walked through the streets of Belfast to the delight of the crowds. Having just taken the bold step of pledging US troops for Bosnia, both Clinton's affirmation of commitment to the Atlantic Alliance and his exhortations to 'take risks for peace' in

Northern Ireland carried renewed credibility. The US also played a key role in the transformation of NATO which, in September 1995, published its 'Study on NATO Enlargement'. This examined the 'how' and 'why' of enlargement, calling for the extension of full security guarantees and all other NATO benefits to new members, but without saying 'who' and 'when'.

US enthusiasm for enlargement seemed to lose steam as it recognised the costs involved, as well as hearing Russian complaints, but the US remained committed to taking in new members. This would inevitably cause problems with Russia, which the US desperately wanted to keep on the path to economic and political reform, a goal that threatened to slip away. US hopes for Russia seemed to focus on the idea that current President Boris Yeltsin would still be in power after the mid-1996 presidential elections.

US efforts to lead in Asia in 1995 were less successful than elsewhere. Hoping to put questions about US engagement behind it, the Department of Defense released its 'Strategy for the East Asia-Pacific Region' in February 1995, which called for 'deep engagement' and reiterated the US commitment to keep 100,000 troops deployed in the region. But the effort to demonstrate commitment and smooth relations with Japan were soon derailed. Still trying to pry open the Japanese market for US cars and car parts, the US on 17 May threatened 100% tariffs on Japanese luxury cars. Unlike in other negotiations, the Japanese, facing serious economic problems of their own, were in no mood to accommodate, and a mini-trade-war was only narrowly averted on 29 June by a last-minute deal in which both sides claimed victory, but which in fact was a US climbdown.

Even worse for US–Japan relations, on 4 September three US soldiers based on Okinawa raped a young Japanese girl, a crime that led to massive protests against the very US presence that Washington was trying to maintain. And that military presence seemed all the more necessary as China began to assert its power in the region and US relations with Beijing deteriorated. In June, and under pressure from Congress, the administration granted a tourist visa to Taiwan's President Lee Teng-hui, provoking a vehement reaction from Beijing, including denunciations of US betrayal and military threats against Taiwan. When China tried to affect the Taiwanese presidential elections in March 1996 by mounting massive and threatening manoeuvres, the US sent two aircraft carrier battle groups to the area to 'monitor' the situation. Even so, US officials remained reluctant to talk about 'containing' China and preferred to emphasise their strategy of 'engagement'. But with questions arising about the continued US military presence in Japan, and an unwillingness in other Asian countries to ally explicitly with the United States, that problem could not have been too far from officials' minds. US relations with China thus worsened, but relations with another old Asian adversary improved: on 12 July, with

widespread support from the US people, President Clinton formally recognised Vietnam.

All in all, 1995 was a good year for US leadership abroad. The
successes in Bosnia, Europe and the Middle East were tentative, and the
major issues of relations with Russia, China and Japan threatened to
explode at any time. But with the domestic initiative taken away from
him, President Clinton focused more time and attention on foreign policy,
and found that the US had a critical role to play in the world after all.

Defence Policy

Throughout 1995, the Republican-controlled Congress sought to challenge several key components of the Clinton administration's defence
policy. Accusing Clinton of neglecting US vital interests, Congressional
conservatives sought to enact legislation to raise US defence spending,
prohibit US troops from serving under UN command, reduce the level of
US payments to UN peacekeeping funds and, perhaps most insistently,
build and deploy a national ballistic-missile defence. While Clinton's defence budget request included funding for theatre missile defence, and
research and development for the national defence system, conservative
Republicans called explicitly for a multi-site, national missile defence to
be deployed by 2003. Clinton argued that such a decision would be
premature and would require the United States' unilateral abrogation of
the 1972 Anti-Ballistic Missile (ABM) Treaty with Russia, which it was not
prepared to do. Meanwhile, Clinton denounced the provisions concerning the UN as a gross infringement of his constitutionally granted powers
as Commander-in-Chief.

The two sides battled over these provisions in a debate on the National
Security Revitalization Act, which passed the House in January 1996, but
received a cold reception in the Senate. When the House effort failed,
Republicans tried attaching their favoured provisions as riders to the
$271bn defence budget authorisation. The President vetoed this as part of
the December budget battle. Besides ABM and peacekeeping issues, the
President also cited other reasons for vetoing the defence bill, including a
requirement that the Department of Defense seek supplemental appro-
priations from Congress whenever US forces undertook an unforeseen
operation (such as in Haiti or Bosnia), provisions that would forcibly
discharge HIV-positive soldiers, and a ban on abortions at military installations. In January 1996, after Congress gave way on missile defence,
peacekeeping and contingency funding provisions, the President agreed to
sign the bill into law, leaving the social provisions to take effect. The law
included funding for national missile defence, but no commitment to deploy.

To counter Republican rhetoric about a 'hollow force', the administration, late in 1994, moved $30bn to operational funds in the Future Years

Defense Program (FYDP). Readiness funding, however, came at the expense of procuring and modernising the services' weapons programmes. Seeing an opportunity to score political points and fund some of their pet projects, Congressional Republicans added more than $7bn to the Pentagon's own budget request, $5bn of which went to purchase additional Navy and Air Force fighter planes and an additional amphibious assault ship. Some members of Congress, especially in the House, wanted to increase spending by as much as $3bn more.

Much of Congress' demand for increased defence spending was politics, not strategy; defence spending means jobs, and it also means looking tough in an election year. Yet, while Congress might have been approaching it the wrong way, there was something in its criticisms. Many analysts, even non-partisan ones, felt that the administration's defence strategy was severely underfunded. They doubted that the armed forces could fight two major regional contingencies simultaneously, as called for in the still-valid 1993 *Bottom-Up Review*.

This debate focused on the two-war strategy. Some administration officials, pointing to the example of October 1994 when Saddam Hussein moved troops to the Kuwaiti border while the North Korean crisis flared, argued that it was still necessary. Others claimed it was both unrealistic and too expensive. Amid growing talk of a coming 'defence train wreck' – commitments had to be cut or spending drastically increased – Secretary of Defense Perry hinted in early 1996 that the two-war strategy might be abandoned. Yet he claimed to be still optimistic that the Clinton defence plan could maintain a balance between readiness, modernising current weapons systems, and investing in research and development for the future.

One way to accomplish this task is to find savings in the existing budget. In June 1995, the Defense Base Closure and Realignment Commission recommended that over two dozen installations, including McClellan and Kelly Air Force Bases – two of the Air Force's largest logistics centres – be closed. This would lead to the loss of thousands of civilian jobs in politically important California and infuriated the President. But there was little he could do about it for by law the Commission's recommendations cannot be changed. Perry also called for an additional round of closures in the next few years. In this case Congress may be reluctant to authorise them.

Other attempts to find savings in the budget did not bear fruit. In May 1995, the Commission on Roles and Missions released a disappointing set of recommendations to reduce redundancy among the services. Although supposedly independent, the Commission was heavy with staff detailed from the various armed services and it proved unable to break away from traditional service parochialism. It is true that some savings are expected from previous reforms of the Pentagon's byzantine acquisition system; any hopes, however, that this latest round of proposed efficiency gains from the Pentagon bureaucracy will produce more savings than previous attempts are probably misplaced.

In terms of funding, the Pentagon had mixed success, in some cases more than it wanted. It secured funding for some of its most cherished programmes, but also found itself obliged by Congress to buy platforms it did not ask for. On 3 November, the Defense Acquisition Board approved the final purchase of 80 additional C-17 heavy transport planes, bringing the total Air Force purchase to 120, and leaving the door open for additional orders. By December 1995, C-17s were operating out of Tuzla, Bosnia, and the Air Force was considering an increased per-year purchase which would lower the price of the plane by almost 5%. The F-22 advanced tactical fighter, on the other hand, met increasing development problems as fuel consumption, weight and, consequently, per-unit price all increased.

Although the Department of Defense was against it, Congress voted in favour of full funding for a third *Seawolf* submarine. It claimed that the construction would maintain an indigenous submarine industrial base until procurement could begin on the more versatile New Attack Submarine. Congress also added $450 million to the defence budget to make possible future procurement of B-2 stealth bombers, once the 20 that have already been funded come into service. The Air Force had argued that the money would be better spent in developing advanced precision-guided munitions for the existing strategic bomber fleet, but B-2s are built in California, which has a lot of Congressional votes.

A Shaky Base

Considering the domestic political turmoil that prevailed in the country throughout most of 1995, the US position in the world as the year ended was remarkably strong. It was the undisputed global diplomatic leader and strongest military power and, after two years of uncertainty, the Clinton administration had finally started to assert itself abroad.

Yet complacency about US global leadership would be a terrible mistake. US assertiveness abroad still rested on a fragile base at home, and Americans have still not proven their willingness to back international engagement if its costs begin to rise. As 1996 began, Russia seemed to be heading away from cooperation with the West, the peace in Bosnia was highly unstable, China was threatening its neighbours, and the cease-fire in Northern Ireland was breaking down. It is an open question whether the US, with its cost-cutting and unilateralist Congress, fickle public opinion and forthcoming presidential elections, will be able to provide the steady leadership and enduring commitment that the world clearly still requires.

New Security Issues In Latin America

Leaders in Latin America and the Caribbean looked forward to the post-Cold War world as an opportunity to achieve democracy, economic liberalisation and expanded free trade. In the developing world, Latin America seemed to be the region most affected by these global trends. At the Summit of the Americas meeting, hosted by US President Bill Clinton in Miami in December 1994, the leaders of the region's democracies, which by then included all the American nations except Cuba, triumphantly laid out a framework for regional economic integration by 2005. It proposed to build on the foundations of already existing trading pacts, including the North American Free Trade Agreement (NAFTA), Mercosur, the Andean Pact and the Group of Three (Mexico, Venezuela and Colombia).

With regard to security, however, 1995 was a year of lost opportunity, overcast by the growing awareness that a crop of old and new security issues had replaced the earlier confrontation of the Cold War. Chief among these issues were drug-trafficking, migration, ethnic conflict, continuing hostilities between the US and Cuba, and environmental degradation in one of the earth's most bio-diverse regions, the Amazon Basin, which straddles eight South American nations.

With NAFTA under increasing political assault in the United States, the prospects of expanding it to include Chile, high on the agenda only a year ago, have been indefinitely postponed. Even as US Secretary of State Warren Christopher toured Latin America in February 1996 proclaiming an era of democracy and free trade, US actions there undermined many of his words. In the week that Christopher spoke, US policy towards the region focused on a new confrontation with Cuba and an increasingly acrimonious relationship with Colombia over its anti-drugs policy.

Washington, in 1995, was seeking a sure footing in the face of newly defined security challenges in Latin America, while backtracking on its earlier strong commitment to free trade and regional integration. Thus, in the post-Miami Summit period, security issues returned to the forefront of inter-America relations and national policy. From Guatemala to Chile it was clear that democracy was not yet fully consolidated and that important issues of war and peace remained prominent in the region.

Central America

Guatemala is the only Central American nation not to have concluded a peace agreement with domestic insurgents within the context of the Central American Peace Accords signed in 1986. But during the last two years several partial agreements have been signed between the insurgent Revolutionary National Guatemalan Union (URNG) and the government. These agreements had been mediated with assistance from the United

Nations and the 'friends' of the Guatemalan peace process, including the United States, Norway, Colombia, Venezuela, Spain and Mexico.

The agreements signed were: 'Human Rights' (March 1994); 'Resettlement of Population Groups uprooted by Armed Conflict' (June 1994); and 'Rights and Identity of Indigenous Populations' (March 1995). To verify the accords, the UN established a Verification Mission in Guatemala (MINUGA). Despite these successes, the two sides have still not reached agreement on a permanent cease-fire, originally scheduled for late 1994.

The Guatemalan military seems unwilling to accept the kind of reforms that occurred in El Salvador, including major restructuring of the size, force structure and mission of the armed forces and the dismantling and re-creation of new police forces. The Guatemalan military believes that it won on the battlefield; the negotiations it envisages are over the terms of the insurgents' surrender. The election of President Alvaro Arzu of the conservative *Partido de Avanzada Nacional* on 7 January 1996 over Alfonso Portillo, a candidate backed by former military dictator Efrain Rios Montt, is unlikely to change this basic negotiating position.

Like El Salvador, but in a more modest way, the move towards peace in Guatemala continues to spur some economic expansion. In 1995, its economy grew by 4%, similar to the two preceding years. In El Salvador, the dividend from the Central American peace process was higher. Its economy grew by 6.5% in 1995, slightly higher than in 1994, although less than 8% was recorded in 1993. Yet the transition from civil war to brokered peace has been fraught with difficulties, even though the letter of the agreements has been generally adhered to. Most of the military and police units have been reduced and demobilised, allowing all but a handful of UN observers to leave at the end of April 1996. In the first months of 1996 there were only 90 UN officials left, down from a high in 1992 of more than 1,000 police trainers, military observers, human-rights monitors and electoral experts.

There has, however, been less progress on land transfers in El Salvador, and the *Farabundo Martí* National Liberation Front guerrillas have not managed to transform themselves into a cohesive political party capable of challenging national power. The threat in El Salvador, however, is not that of renewed civil war. It is of social breakdown. Many armed groups, whose ranks are mostly drawn from former combatants on both sides, have sprung up and are now living off banditry and crime. The security issue in 1995 was no longer political violence and national security; it became personal security and social violence. Even as the war ended, the killing did not stop. The Salvadoran Attorney General's office reported that in 1995 there were 7,877 deaths due to violent crime, a homicide rate comparable to the annual death toll during the 1989–92 civil war. This is inordinately high for a country with a population of only 7 million.

In Nicaragua, too, the post-war period has led to an upsurge in armed groups and social crime. The armed forces have been radically reduced, and control over them has been transferred from the Sandinista Party to the Nicaraguan state. This was underscored on 21 February 1995 when long-time Sandinista Army Commander Humberto Ortega finally relinquished control.

Five years in opposition has taken its toll on the entire Sandinista leadership. In January 1995, it split into two factions: the 'renewalist'; and the 'orthodox'. The first group is led by former Vice-President Sergio Ramirez, a well-known poet and novelist. The orthodox wing is led by former President Daniel Ortega. At the same time, President Violetta Chamorro has lost much of the partisan support that led to her election in 1990. With the unravelling of the broad coalitions that had previously polarised Nicaraguan politics, a centrist block has emerged in the legislature. It groups moderate Sandinistas and other parties, and is strong enough to challenge and block both the Chamorro government and the orthodox Sandinistas.

Despite its strength in the legislature, the presidential elections scheduled for 20 October 1996 are likely to be between the hardline Sandinista Daniel Ortega, and the right-wing Mayor of Managua, Arnaldo Aleman Lacayo. Aleman Lacayo has consistently topped the opinion polls and may well thwart the Sandinistas' hopes of returning to power in 1996. Much depends on the economy. It has begun to grow by a modest 1–2% during the last two years, but the nation has still not recovered from the economic contraction of the war years in the 1980s. In a time of fiscal austerity, US leaders have been reluctant to provide large amounts of foreign aid to Nicaragua, as much of the population once expected when they voted the Sandinistas out of office in 1990.

Panama's faltering economy and weak institutions have surprisingly led to a new national debate over the fate of the Panama Canal. According to the 1977 Torrijos–Carter agreements, the Canal will revert to full Panamanian control at midnight on 31 December 1999, ending the US presence in the country. The US Southern Command, which has become a front-line base in the post-Cold War world and a major actor in the hemisphere-wide war on drugs, is to relocate to Miami. President Ernesto Perez-Balladar of the ruling Revolutionary Democratic Party (the party of Generals Omar Torrijos and Manuel Noriega) has ruled out renegotiating the original treaty. But he has indicated that his government would be receptive to new negotiations leading to a continued US presence. Such an action would go totally against Panamanian public opinion.

Panama continues to be implicated as a major transhipment and money-laundering centre for the Andean drugs trade. In the last year, coca fields and cocaine laboratories were set up for the first time inside the country near the southern border with Colombia. While Panama is increasingly integrated with its Andean neighbours in the illicit narcotics trade, it is also making overtures to join the Andean Pact trading agreement formally.

The Andean Region

Renewed post-Cold War conflict in the Andean region is reflected in the rise of drug-trafficking and the corresponding anti-narcotics war, mostly defined by a web of bilateral relations between individual countries and the US. In the central Andean countries of Colombia, Peru and Bolivia, the anti-drug war overshadows most other issues in diplomatic and political relationships with the United States.

The drug trade, with its booming export profits, and the drug war, with its broadened military and police infrastructure, have had an inordinate impact on the shape of politics in the region. This was underscored in 1995 in Colombia where President Ernesto Samper has been enmeshed in a major corruption scandal concerning the alleged financing of his 1994 presidential campaign by the Cali drug cartel.

The drugs trade has led to a degree of economic integration in the Andean region with fields in Bolivia and Peru supplying production facilities in Colombia. By the mid-1990s, however, its geography had considerably altered. Repressive anti-narcotics policies pushed production well beyond its original epicentre in the Andes. Within the principal Andean drug-producing nations production has spread across a larger geographic area and, more seriously, crops and laboratories are now appearing in Brazil, Paraguay, Ecuador and Panama.

In Colombia, President Samper came to power as a progressive reformer who vowed to negotiate a peace settlement with the nation's guerrillas and to sign international human-rights agreements to strengthen the nation's capacity to confront major human-rights abuses by its security forces. On taking office in August 1994, Samper also vowed to implement a 'social pact' to ease the conditions of those displaced by Colombia's integration with the global economy.

However, by mid-1995, after only one year in office, he was deep in political scandal as evidence surfaced that his campaign had accepted money from the Cali cartel in the closing days of the 1994 presidential race. The scandal led to the resignation and imprisonment of Samper's former campaign manager and subsequent Minister of Defence, Fernando Botero. The crisis boiled over when Botero, after six months in prison, changed his story and directly accused the President of taking the money and orchestrating a massive cover-up. Samper insisted that he was innocent, and pointed to his administration's aggressive anti-narcotics strategy that included imprisoning the principal leaders of the Cali cartel and establishing a major programme of crop eradication that had eliminated half of the known coca, opium poppy and marijuana fields, totalling 30,476 hectares.

On 1 March 1996, the US compounded Samper's troubles by declaring that Colombia was no longer cooperating in the war against drugs. The US announced that it would not certify Colombia, as required by US law, for continued support of international loans. In taking this step the US had branded Colombia a pariah nation, in the same category as Iraq, Syria and Myanmar. US officials were careful to state that their intent was not to damage the Colombian economy or to hurt its citizens; the de-certification was to express its official disapproval of Colombia's President.

Most analysts believe that these actions will not seriously damage Colombia's booming economy, although there may be regional repercussions on certain products. Even without considering the illegal export booms in cocaine, heroin and marijuana, Colombia still has a very strong economy, built on a solid foundation of legal exports and domestic manufacturing. It has only recently begun to enjoy the benefits of exporting petroleum from major fields discovered in the eastern plains. In the last two years, Colombia has experienced 5.8% and 5.2% growth on top of more than 40 years of economic expansion without recession.

While bilateral relations between the US and Colombia hit rock bottom in 1995, Colombia is more preoccupied with its presidential crisis. In March 1996, a Colombian Congressional committee conducted hearings, and a vote on whether to recommend formal impeachment proceedings was expected. Everything else is currently on hold. Samper's social pact, the peace process with the guerrillas, and human-rights efforts have all been marginalised as his power has diminished. If Samper does survive politically, he is unlikely to recover his lost authority. If he steps aside or is impeached, he is likely to be replaced by Vice-President Humberto de la Calle. For Colombia, the US decertification process has acquired a bitter taste, proving, as several Colombian officials have expressed it, that the US is an unreliable friend and ally.

Venezuela is undergoing a different kind of political instability. President Rafael Caldera presided over a mismanaged and contracting economy in 1995 and has retained little political support. After two years of negative growth in 1993 and 1994, the economy finally grew by 2%. But this hardly counted with the people and, in April 1995, workers, teachers, students and neighbourhood groups took to the streets demanding improved social services. Caldera vowed to ignore the International Monetary Fund (IMF's) rules for disbursing funds. By early 1996, however, he desperately needed cash. At that point he backed down and entered into serious negotiations with the Fund.

As presidential authority diminished, the left accused Caldera of planning an *autogolpe*, or 'self-coup', that would close Congress and concentrate power in the executive, as Peru's Alberto Fujimori had done in 1992. In the regional elections in December 1995, Caldera's supporters were thoroughly trounced. In the oil-rich state of Zulia, Lieutenant-Colonel Hugo Chávez Frías, the former Venezuelan Army commander who had led an aborted coup on 4 February 1992, was elected governor with the support of the left.

In the face of such political and economic instability, Caldera turned to a familiar sector to revive his and his country's fortunes: oil. Twenty years after foreign oil companies had been nationalised, Venezuela welcomed back several large multinationals, including Mobil, Nippon Oil, BP and Amoco Conoco. The plan was to double production over ten years, essentially abrogating prior Organisation of Petroleum Exporting Countries (OPEC) commitments. New oilfields are being exploited on the border with neighbouring Colombia, with drilling equipment in at least one oilfield literally yards from the border. Colombia has protested at this. Together with Colombian guerrilla incursions into Venezuelan territory, border tensions remain on the top of the agenda for these two Andean countries.

Ecuador has been suffering the financial and political aftershocks of its border war with Peru which began on 11 January 1995. The guarantors of the original 1941 protocol establishing the borders (the US, Brazil, Chile and Argentina) stepped in and imposed a cease-fire which took effect on 11 February with both sides declaring victory. The war seriously undermined the Ecuadorian economy and forced the government to impose emergency taxes, and economic growth stalled after two previous years of robust growth; the boost that President Sixto Durán derived from his nationalist stance at the time of the war proved short-lived. Furthermore, Vice-President Alberto Dahik was forced to resign over corruption charges, and he fled the country rather than go to prison.

Presidential and legislative elections in Ecuador are scheduled for May 1996. Leading in the polls is Conservative Jaime Nebot. However, Freddy Ehlers, a television presenter, taking advantage of the new electoral laws permitting individuals to run without a party, announced that he will run for president. He immediately moved into second place in the polls and has gained the support of Ecuador's increasingly active and strong indigenous movements. The leaders of Ecuador's indigenous confederation, *Confederación de Nacionalidades Indígenas del Ecuador*, also announced that they will put forward candidates for Congress.

The emergence of the Indian community as a strong political actor within the system marks a major change in Ecuadorian politics. This reflects region-wide trends throughout the Andes, particularly Bolivia – where an Aymara Indian, Victor Hugo Cardenas, was elected Vice-President in 1994 – and in Colombia, where special electoral districts were created in 1991 for the indigenous community guaranteeing representation in Congress. The political incorporation of indigenous communities occurred after a decade of increasing protest and violence, and their assimilation is still one of the most explosive and potentially violent social issues on the national and international agendas.

In Ecuador, the indigenous communities of the eastern rain forests have sued the Texaco Oil Corporation in a New York court for the environmental destruction of their homeland. In May, they rejected a deal struck by Texaco and the government to clean up the environmental damage. Mobilising their followers, the national and regional federations demanded a better deal.

Peruvian politics were dominated in 1995 by the successful re-election of Alberto Fujimori on 9 April, under a new Constitution that he had created to allow him to run for office again. The border war with Ecuador had only a minimal impact on the Peruvian economy, although it may have aided Fujimori's re-election bid. In contrast to Ecuador's Sixto Durán, who was only nominally in control of the armed forces, President Fujimori worked closely with the military throughout the conflict.

Fujimori's electoral triumph was ensured by economic growth and the defeat of the guerrilla insurgencies. In 1994, Peru led the world with almost 13% growth. In 1995, expansion still continued at almost 7%. Moreover, Fujimori had forced two large guerrilla organisations, *Sendero Luminoso* (Shining Path) and the *Tupac Amaru* Revolutionary Movement, to their knees. His opponent in the presidential context, former UN Secretary-General Javier Perez de Cuellar, was reduced to complaining that Fujimori was not democratic, while acknowledging the President's remarkable successes. Fujimori won the election with over 65% of the vote.

Despite allegations of high-level government and military involvement in drug-trafficking, Fujimori has mended relations with the US. In the last two years, the Clinton administration has certified Peru as officially cooperating in the war on drugs. Nevertheless, Peru still remains the principal supplier of coca and coca paste, the refined product that is subsequently converted into cocaine.

Southern Cone and Brazil

In Argentina, as in Peru, the national Constitution was changed to permit the re-election of a popular president. On 14 May 1995, Carlos Menem won his bid for re-election by 49.5% and gained a majority for his Peronist *Partido Justicialista* in the legislature. Unlike his Peruvian counterpart, he managed this during a period of economic downturn, although one that followed several years of steady economic growth and low inflation. On the eve of the national elections, when unemployment had reached 14%, Menem was forced to announce austerity measures. It appears, however, that voters feared the return of a hyper-inflationary economy without Menem's steadying influence and they returned him to office on this basis.

Menem must still confront issues of human-rights abuses by the armed forces during the dirty war of the 1970s and 1980s. In May 1995, two high-ranking officers publicly admitted participating in atrocities. Menem took a conciliatory tone, insisting that the country should look to the future and not the past. Yet he has also made great strides in radically restructuring the nation's armed forces, and in civil–military relations.

Since taking office, Menem has sharply reduced the size of the armed forces, halving the number of generals and 80% of the troops, leaving just 20,000. Defence spending during the last decade has been slashed by 75%. The President has also overseen the creation of new military missions, and Argentina now leads Latin America in participating in peacekeeping operations. It recent years it has sent, or will soon deploy, peacekeeping forces to Bosnia-Herzegovina, Haiti, Kuwait, Mozambique, Angola and the Western Sahara.

Menem's policy of *rapprochement* with the UK over the Falkland/ Malvinas Islands is a clear sign of the change in civil–military relations. Having re-established diplomatic relations in 1990, the President welcomed UK foreign investment into Argentina's liberalising economy. On 27 September 1995, Menem signed an agreement with the UK permitting joint oil operations around the disputed islands. The agreements are still vaguely worded, but they open the way for exploration and eventual exploitation by both the UK and Argentina of what are believed to be large oilfields. Nevertheless, the issue of sovereignty over the disputed islands still disturbs relations. To underscore this point, the UK announced on 3 March 1996 that its naval vessels will begin patrolling the waters off the island of South Georgia to prevent unauthorised fishing by Argentina in the disputed waters.

In Chile, President Eduardo Frei has also had to confront the issue of civil–military relations during the past year. On 30 May 1995, the Supreme Court confirmed a seven-year sentence against former Security Chief Manuel Contreras who was convicted of ordering the assassination of Allende government Foreign Minister Orlando Letelier in Washington in 1974. The former Chilean strong-man, General Augusto Pinochet, who is still Commander-in-Chief of the Armed Forces, prevented Contreras' arrest for several months. In November 1995, Pinochet arranged for him to be incarcerated in a special, more comfortable facility.

As the Contreras incident demonstrates, Chile has still not put the issue of past human-rights abuses behind it. Trials of other former military officials continue. Moreover, the incident highlights the continued limits placed on civilian authority, and the absence of a clearly defined relationship between civilian and military power, particularly over past human-rights abuses. The resentments of the past will be difficult to overcome, but some avenues of possible improvement are being explored. Chile has also begun to play an active role in world peacekeeping, sending troops to Kuwait, Cambodia, Israel and Kashmir. Such missions may help to define the identity and position of the armed forces in the post-Pinochet era.

In Brazil, President Fernando Henrique Cardoso, who has faced much opposition from the social and business sectors, has been able to steer a steady economic course, ensuring growth of 4% in 1995, following 7%

growth in 1994. Inflation is currently falling, reserves are at an all-time high and recession, much feared in some labour and business quarters, has been averted.

Brazil is distinguished from its Southern Cone neighbours by having no lingering human-rights issues left over from its long history of military rule. Although they retained certain prerogatives after handing government back to civilians in 1985, the armed forces no longer present a major challenge to democratic rule.

One high priority during the military era, however, continues to be a source of conflict today. This is the issue of Brazilian control over its Amazon territory, which occupies over 58% of its national territory (5.2 million square miles). With Brazil's growth as a producing, transhipment and consuming nation in the global drug trade, control over remote areas within its national territory has become an even more pressing security issue. This is compounded because the Amazon Basin has become a high-profile issue for international environmental, human-rights and indigenous-rights groups.

The confluence of national and international interests over the Amazon led to a major government scandal in 1995. President Cardoso had announced plans to create an Amazon Monitoring System (SIVAM), using land- and space-based radars, which would allow the government to monitor drug-trafficking, environmental change and possibly mineral deposits in areas that have remained beyond the state's direct influence or control. In May 1995, the government awarded the SIVAM contract to Raytheon, a US company.

This caused an uproar, with accusations of US intervention in internal security affairs, to which were added charges of financial manipulation and influence peddling. Indeed, the role of SIVAM stimulated a political debate over national sovereignty, security policy, environmental rights and the need for such a system in pursuing different security and environmental ends. President Cardoso was forced to fire one of his closest aides for influence peddling. By March 1996, the Senate still had not approved the deal, although it is expected to do so later in the year. The Amazon is destined to play a heightened role in Brazil's security and development plans however the issue is resolved.

The Caribbean

Historically, the Caribbean has been the site of US gunboat diplomacy and unilateral interventionism. In 1995, US security interests continued to be challenged in this often-neglected and misunderstood region of the world.

The US intervention in Haiti in September 1994 restored President Jean-Bertrand Aristide to power on 15 October 1994. As the US had advertised, its troops handed authority over to UN multilateral peacekeeping forces on 25 March 1995, and then most left the island. The UN and the Organisation of American States (OAS) oversaw municipal and legislative elections in June, and on 17 December 1995 presidential elections were held to replace Aristide whose constitutional term had expired; Haiti's Constitution prohibited his re-election. Despite overwhelming popular support and the fact that most of his presidential term had been spent in exile, Aristide eventually agreed to support a successor, René Préval, who was duly elected. Power was transferred on 7 February 1996, marking the first ever successful transition of presidential power from one democratically elected leader to another in Haitian history. Aristide's Lavalas Party continues to control the executive branch and the majority in both houses of Congress.

In his short stint in office, Aristide tried to effect measures that would have a lasting effect on Haitian life. On 1 May 1995, he announced that he had demobilised and abolished the Haitian armed forces. He also instituted a multi-phase, five-year programme to train a new, professional police force drawn from a wide spectrum of the populace and trained in police activities and respect for civil and human rights. An interim force was established during 1995.

Haiti's economy, however, remains weak. In the 1980s, it contracted by more than 2% annually. Under the military government, economic activity fell by over 30%, unemployment rose to 70–80% and per capita income remained the lowest in the western hemisphere at about $250. For a short period immediately following Aristide's return, Haitians remained at home in the hope of economic improvement, but they are now once again taking to the sea. Their desired destination is the US, but many are turned back by the US Coast Guard.

The UN peacekeeping force in Haiti was scheduled to leave at the end of February 1996. But both Aristide and President Préval, fearing instability following the withdrawal, requested an extension of the UN mandate on 5 January 1996. The UN agreed in principle to a four-month extension for a reduced force, but was unable to raise the necessary funds for any but two battalions of troops from Bangladesh and Pakistan, along with an Argentine field hospital. The Canadian government stepped in and agreed to pay for 700 of its own troops under the command of a Canadian general, and 100 Royal Canadian Mounted Police who will train a new police force.

The other area of the Caribbean with which the US remains deeply involved is the neighbouring island of Cuba. During 1995, the US and Cuba kept up their mini-Cold War, seemingly divorced from the logic of larger global change. Cuba has suffered greatly from the collapse of the Soviet Union. Its economic dependence on the former communist superpower was underscored as its economy contracted by 45% between 1991 and 1994. Most analysts agree that the economy hit rock bottom in 1994 and that it has begun a slow recovery in 1995 when GDP grew at an estimated by 2%.

President Fidel Castro's government has openly solicited foreign investment, particularly in the tourist, mining, pharmaceutical and bio-medical sectors. In September 1995, Cuba's National Assembly passed a new foreign-investment statute opening most areas of the economy and guaranteeing full ownership, profit remittances, infrastructure support and other provisions for a stable investment climate. Cuba had already attracted 212 joint ventures worth $2bn with companies from over 50 countries. At the same time, the government legalised possession of US dollars and opened up certain areas of economic activity to the market. Most notable – and successful – was the market activity of a reformed agricultural sector which immediately helped ease the pronounced, so-cially destabilising food shortages. The government also introduced its first income tax to channel revenues from these market activities to the state.

Cuba continues to insist that its economy is being strangled by the US economic embargo. Nevertheless, Cuban officials claim that their principal foreign-policy goal is to normalise relations with the US without, in their words, sacrificing their national dignity or the achievements of the Cuban revolution. In 1994, the two countries managed to reach agreement on immigration which effectively eliminated the special status granted to illegal Cuban immigrants previously regarded by the US as political refu-gees. In October 1995, President Clinton loosened some of the sanctions implemented in August 1994 at the height of the migration crisis. This appeared to be a modest first step, that could be expanded after the US elections in November 1996.

That promise, however, was shattered when Cuban authorities shot down two unarmed civilian aircraft on 24 February 1996. The planes were flown by exiles belonging to a Miami-based group dedicated to over-throwing the Castro government. This group originally monitored the waters between Cuba and the US to aid those trying to escape on rafts and small boats. As the numbers of would-be escapees dropped, the group began to fly over Cuban territory to drop leaflets instead. On this occasion, only one of the three planes transgressed Cuban airspace, but it was the other two that the Cuban Air Force shot down.

This incident once again sent US–Cuban relations into a severe de-cline. In retaliation, President Clinton reversed course and endorsed legis-lation that had been delayed in Congress to tighten the US embargo further and place sanctions on third-country investors with interests in both Cuba and the US. The bill is aimed at reducing the growing interna-tional investments that are easing Cuba's economic crisis. As UN General

Assembly resolutions condemning the US embargo demonstrate, the United States is increasingly alone on this issue. In 1995, only the US and Israel voted for UN-imposed sanctions against Cuba. Moreover, the US has angered some of its principal trading partners, most notably Canada, over recent legislation tightening its unilateral embargo.

Facing New Security Issues

Change and conflict continued to characterise developments in Latin America in 1995. Although a return to democratic government and a strong push towards regional integration defined the political and inter-national agendas in the years immediately following the end of the Cold War, security issues have moved to the forefront of regional govern-ments' concerns. These ranged from the ongoing conflict between the US and Cuba, US concern over instability and migration flows in the Carib-bean, the US-sponsored war on drugs which is creating a new infrastruc-ture for its military involvement throughout the region, the rise of new indigenous actors in national politics – particularly in Guatemala and the Andes – and the redefining of environmental issues as security issues as in Brazil. All these issues are potentially destabilising, and could draw in outside powers. In this region, as in others, 1995 has confirmed that the post-Cold War world may be a more plural world, but it has still not become a safer or more stable world.

Mexico: The Legacy Of Reform

In both economic and political terms, 1995 was a tough year for Mexico, which lost the political stability that had characterised its recent history and distinguished it from so many other Latin American countries. The conventions of the ruling political class were challenged and party disci-pline weakened. A decade of economic adjustment had taken its toll on the rules and institutions which, in the past, had regulated the political sys-tem. The ruling party's hegemony has declined while opposition parties gradually consolidated. These shifts, however, may well lead to a more stable Mexico.

The legacy of former President Carlos Salinas' administration is a mixed one. Mexico's economy has remained particularly sensitive to po-litical developments, but by the end of 1995 there were clear signs of adjustment and recovery. Similarly, while the peasant uprising in Chiapas revealed the vulnerable sides of Salinas' economic and political reform, the government managed to establish a cease-fire and to enter into negotia-tions with the guerrillas. While the opposition parties were effectively getting their act together, the ruling party also showed that it could adjust. Mexico's political and economic life now appears on the path to a new and healthier equilibrium than in previous years.

Shifting Political Realities

Although the August 1994 presidential elections did not fulfil the hopes of those who saw it as Mexico's first post-authoritarian election, the contest was typical of many 'first elections': relative uncertainty underlay both the results and their acceptance by all the contenders; there was a high turn-out; and it involved an inflated number of parties. The elections did not, however, replace the Institutional Revolution Party (PRI) which has ruled Mexico without interruption since 1929.

That political change is indeed under way in Mexico, despite the PRI's electoral success, was shown by the fact that it was returned to office with a much lower majority than usual. Newly elected President Ernesto Zedillo, who is more a bureaucrat than a politician, has recognised that these elections were clean, but not wholly fair. This has strengthened the position of those who insisted that the 1994 elections were not governed by rules that could ensure fairness and equality, or that could guarantee all parties an equal chance of gaining power.

The opposition may have been unable to win the major prize in 1994, but since then it has achieved a vigorous and legitimate presence in the political system and now offers electors 'options of voice and exit'. Former President Salinas' efforts to regulate the pace of political change through electoral engineering were overtaken in 1995 by the opposition which successfully created a new framework for party competition. The party system is now characterised by competition and cooperation among the three main contenders: the PRI, the National Action Party (PAN) and the Democratic Revolution Party (PRD).

Since the 1930s, the PRI has played a key role in Mexican politics. It demilitarised political competition and offered a solution to peaceful succession. Not only did the Party widen the regime's social bases, but it also became the main arena for political competition. Since the 1980s, however, its capacity to ensure peaceful succession has been questioned. It has been forced to compete and to share power with other opposition parties, a process that exacerbated the already fierce competition for PRI electoral posts and further contributed to internal divisions.

Zedillo, like his predecessor, was forced during the electoral campaign to commit himself to party reform. His plans to reshape relations between party and state and to distance himself from the PRI's internal strife has created considerable confusion within the Party's ranks. Clearly, Zedillo's reforms were a response both to the demands voiced by an opposition that now holds significant power, and to the new role played by the presidency as mediator between the governing party and the opposition. Although over the past year Zedillo has been forced to accept the need for PRI leadership, the new role of opposition parties in maintaining political stability has reduced the likelihood of a smooth return to the traditionally harmonious relationship between the presidency and the PRI.

The electorate's determination to proceed with ordered and peaceful political change has been demonstrated by accelerated political liberalisation since the mid-1980s. The first opposition government was elected at the state level in 1989. By 1995, four of the 32 states were in the hands of the opposition, as were 300 out of a total of 2,500 municipal governments. Although other forms of political participation, including mobilisation, have continued, elections have now become the dominant pattern. This was reflected in the 1994 election, both by the high turnout (77.8% of all registered voters cast their vote) and by the large number of voluntary electoral observers (81,620). Moreover, Mexico's economic crisis acted as a catalyst in shifting the electorate's perception of the opposition's capacity to govern.

The PAN, which some considered an extreme right-wing organisation, has, like Mexico's other political parties, evolved. In governing increasingly large areas, it has developed traditional left–centre–right divisions. In 1987, only 800,000 Mexicans were under PAN rule; by the end of 1995, the party held the governorship of four states (Baja California, Guanajuato, Chihuahua and Jalisco) and controlled 222 municipal governments, including 13 regional capitals or important urban centres.

The PAN's return to power for a second term in the states of Baja California and Guanajuato in the November 1995 elections was a measure of the electorate's renewed trust. Those elections also gave the party a foothold in territories which had traditionally been the stronghold of the PRI or the left-wing PRD. Despite these impressive advances, the PAN has demonstrated a limited capacity to formulate coherent economic and foreign-policy programmes – which can be partially explained by its need to concentrate on fighting the PRI's biased electoral practices.

The left-wing PRD had been the second largest electoral force in 1988, but it lost this position to the PAN three years later. Although the 1994 electoral results confirmed this trend, the PRD maintained its almost absolute electoral dominance over other left-wing parties. This dominance may be coming to an end, however. Over the past six years, the PRD has been at war with itself and paralysed by dual leadership. Unless it can overcome enduring divisions, the position of the party, as well as its overall leadership over social movements and left-wing groups, will deteriorate even further.

This has been clearly seen with regard to the *Ejército Zapatista de Liberación Nacional* (EZLN) guerrilla movement in the southern state of

Chiapas, whose relations with the PRD have not been particularly smooth. In May 1994, *Subcommandante* Marcos, leader of the EZLN, responded to efforts by the PRD presidential candidate Cuauhtémoc Cárdenas to strengthen the links between the two groups by publicly humiliating him. By holding out for a more radical position the rebels may even bear some responsibility for the poor performance of the left-wing parties in the 1994 and more recent elections.

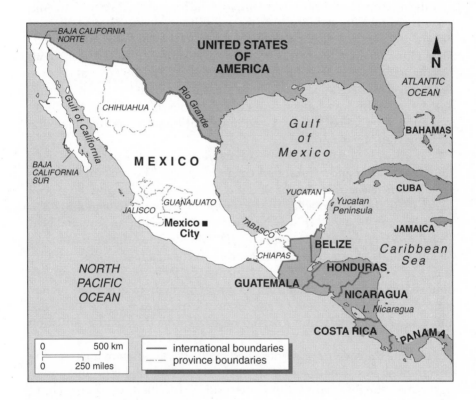

An Economy Under Siege

The devaluation of the peso in late 1994 was the first of a number of events that threw a harsh light on the success of economic and political reform under the Salinas administration. The former President's economic pro-gramme, characterised by rapid trade liberalisation leading to the North American Free Trade Agreement, the privatisation of state-owned compa-nies and the deregulation of the financial markets, had mixed results. Although the December 1994 financial crisis unleashed a heated debate, the hard statistics were not too serious. The overall balance included a 3.0% growth in real GDP, a 15.9% inflation rate, a 17.7% gross national savings/GDP rate, a public-sector deficit of -2.8% of GDP, and a current-account deficit of -5.6% of GDP. One of the main achievements of Salinas' economic policies was to reduce both external and internal debt. External obligations were decreased under negotiations carried out within the framework of the Brady Plan, while internal debt reduction was financed with revenues from the privatisation process (approximately 10% of GDP).

President Ernesto Zedillo took office on 1 December 1994, having won Mexico's first relatively free election in which votes were properly counted and the electoral authorities were impartial. This electoral result had suggested a return to stability, but the magnitude of the current-account deficit, renewed political instability and financial mismanagement undermined the sustainability of the prevailing exchange rate. Reserves fell with no compensating contraction of monetary policy. The US Federal Reserve fund rate increased. Mexico sought to finance its deficit by dollar-denominated short-term investments in public debt bonds (*tesobonos*) which by January 1995 totalled $25bn. A rise in capital flight, however, made the exchange rate unsustainable.

Shortly after Zedillo took office a severe attack on the peso triggered a 15% devaluation and a profound economic crisis. The scale of the crisis soon became clear as central bank reserves evaporated, leaving the official parity extremely vulnerable. The provision of an unprecedented $17.8bn stand-by loan by the IMF was justified on the basis of an 'unparalleled global financial crisis'. Since January 1995 the Zedillo administration has devoted all its efforts to restoring conditions for macroeconomic stability and renewed economic growth. In January 1995, the government announced a programme featuring a slower growth rate for the money supply, a drastic reduction in government expenditure, as well as price and wage-control measures to avoid returning to a devaluation–inflation cycle.

Despite the severe crisis unleashed by the devaluation of the peso, the government was able to negotiate with business and labour groups and to sign a new stabilisation pact in January 1995. The plan relied on a credit line of $18bn from the United States. Prospects of strong opposition within the US Congress forced President Clinton to withdraw the original proposal from Congress and to resort to the Treasury's Exchange Stabilization Fund instead. The delay and difficulties surrounding the negotiations for external support added to the instability of Mexico's financial and exchange markets.

Expecting a dramatic drop in the Mexican stock market and that Congress might attach unacceptable political conditions to the plan, Clinton bypassed Congress, announcing on 31 January that he was establishing a $50bn rescue plan. The new package would underwrite an estimated $14bn Mexican current-account deficit for 1995, and *tesobonos* worth

$29.2bn. The new plan significantly extended the repayment period from a few months to three to ten years.

Clinton's rescue plan may have momentarily avoided a costly dispute with Congress, but it transferred the confrontation to the Group of Seven summit in Toronto on 15–17 June where concern over the IMF's liquidity and monitoring capacity was voiced. To meet these objections, the US Secretary of the Treasury and the Mexican Secretary of Finance and Public Credit entered into four agreements for the credit, but with strict conditions. These included careful transparency, stringent financial reporting requirements, and an 'Oil Proceed Facility Agreement' under which buyers of Mexican oil were to send payments to a special account in the Federal Reserve Bank in New York to ensure that Mexico did not attempt to stop its loan repayments. Mexico will be able to draw on this account if it complies with the accord's conditions.

This massive external assistance did not erase all concerns about the viability of Zedillo's economic programme. By early March 1995, the peso had fallen to 7.5 to the dollar, a devaluation of 120% since the President's inauguration, and nearly 23,000 jobs had been lost. On 9 March 1995, a more severe adjustment programme was announced that reinforced the economic programme with additional measures aimed at eliminating the trade deficit and containing the inflationary impact of the peso devaluation. The programme had been designed by the Mexican government, the US Treasury and the IMF, and introduced a number of steps aimed at fostering investors' confidence.

By July there were enough signs of confidence for Mexico to return to the capital markets, and by early October *tesobono* obligations had been reduced from $29bn to $2.5bn. The optimistic view expressed by the IMF *World Economic Outlook* at these developments was clouded, however, both by the magnitude of the recession and by the continuing volatility of the exchange rate. In addition, the economy was buffeted by adverse political developments. The intimate interaction between political events and financial instability was demonstrated by a new run on the peso, unleashed by strong evidence of links between senior members of the PRI establishment and drug-traffickers, as well as by rumours of a military coup.

These rumours were subsequently dispelled, but their impact on the exchange rate made clear that market confidence was weak and that the administration faced great difficulty in restoring economic credibility. Central bank intervention in the two last months of 1995 brought back relative exchange-rate stability, and as the year ended the changes in the political arena also strengthened stability. Both the result of the 12 November 1995 elections, which revealed the strength of the right-wing opposition PAN, the increasing isolation of the guerrillas in Chiapas, and their apparent willingness to negotiate suggested a relatively stable start to 1996.

Clearly, economic recovery and Mexico's longer-term prospects will largely depend on its ability to maintain the reforms that have been set in motion. As expected, 1995 closed with a GDP contraction of 7%, a 45% inflation rate and significant interest-rate fluctuations. Although the threat of *tesobono* insolvency was dissipated, stability has not been fully restored, and given such conditions it is not yet clear whether the 33% growth in exports in the first ten months of 1995 will enable Mexico to meet the external debt obligations of $15bn that are due in 1996.

Revolution and the Peasantry

Unlike traditional insurgent movements, the EZLN has been described as a group of 'post-modern' insurgents who rely more on the power of the media than on military capabilities. The Mexican Army estimates that the movement has only about 1,500 well-armed men, and several thousand poorly armed and poorly trained members. The Mexican military, recognising that the movement is more of a political-military body than a military structure, has accepted the need for social and political answers to the problems in Chiapas. Two years of massive deployments and gradual advance have left the movement effectively encircled and have allowed the military to set in motion counter-insurgency policies. A new military zone was created, 12,000–14,000 troops were deployed and not only the maintenance of public order, but also the provision of minimum services in the area of conflict now depend on the armed forces.

There is little doubt that NAFTA has played a key role in the evolution of the conflict. The fact that NAFTA was nearing a vote in the US Congress led the Salinas government to underplay the insurgency threat, helping the small group to find a foothold. Mexico's armed forces became the target of strong criticism from human-rights organisations, while the Zapatistas took advantage of the unusually high level of international attention to push through their demands. This international pressure has in turn influenced and helped to enforce the EZLN's commitment to negotiations and peaceful solutions.

A cease-fire was established as early as January 1994 and was briefly interrupted only in February 1995. Throughout these two years, the government and the Zapatistas have held successive negotiations. In August 1995, a poll conducted by *Alianza Cívica*, a non-governmental organisation, showed that three million Mexicans favoured the transformation of the insurgent movement into an independent political party. This increased the pressure for an acceptable settlement.

The first results of protracted negotiations came in November 1995 when greater autonomy was granted to Indian communities. The government recognised the communities' traditional forms of social and political organisation and their right to most natural resources, except strategic assets like oil. On 2 January 1996, Marcos celebrated the second anniversary of the rebellion by announcing the decision to create a new *Frente Zapatista de Liberación Nacional* aimed at peacefully amending the Constitution. It may not make much impact, however, for the movement's social base has apparently eroded over the past year and it remains geographically limited to Chiapas. Whilst the role of the army in maintaining public order in the region has awakened significant discontent, its presence there has also gradually fostered dependency links with Indian communities under the EZLN's influence.

Relations with the United States and NAFTA

NAFTA has intensified the interaction between the US and Mexican political systems. An early indication of this was Congress' insistence that Mexico accept parallel agreements on labour and the environment. Although NAFTA did not contemplate creating a supranational institutional framework, nor insist on Mexico's accession to formal democratic commitments, it internationalised Mexican politics.

Since January 1994, when the Chiapas uprising began, the 'splendid isolation' enjoyed by Mexico's regime came to an end. Moreover, to the extent that the economic crisis made clear the need for greater monetary and financial coordination, further asymmetric encroachments on the sovereignty of both the government and the EZLN can be expected. The conditions attached to the US rescue package have already increased its surveillance of Mexico, and the dynamics of drug-trafficking have forced both countries to cooperate in areas ranging from policing the borders and detecting aircraft entering Mexican airspace, to controlling money-laundering. The closer interaction between the two political systems will exert significant pressure on the skills of politicians in both countries. Indeed, not only has the framing, negotiation and subsequent disbursement of the rescue package increased international monitoring of Mexico, but the severe economic crisis has again awakened anti-Mexican feelings in the US.

Although in the context of NAFTA both countries are expected to seek to isolate such difficult issues as drug-trafficking and migration from other overarching bilateral considerations, the potential of these to poison US–Mexican relations is still significant. The political capital that President Clinton invested in NAFTA not only helps explain his administration's continued support of Salinas' candidacy to head the new World Trade Organisation, but also Clinton's apparent decision to pursue a policy of cooperation rather than diplomatic confrontation when dealing with sensitive issues.

Throughout 1995, one of Clinton's major policy objectives was to prevent those intent on discrediting NAFTA from succeeding. This was confirmed during the October summit between the US and Mexican presidents in Washington DC. Yet, as successive events demonstrated, Clinton's ability to pursue such a strategy will depend significantly on his ability to contain pressures in Congress, which are certain to increase as the November 1996 presidential election draws nearer. Indeed, closer interaction between the two political systems has exacerbated the domestic political difficulties faced by both presidents when dealing with each other. The policies carried out by the Zedillo administration to reassure the US can easily become targets for nationalist accusations by domestic opponents.

President Clinton campaigned against California's Proposition 187, which set the strictest regulations of any US state on migration and immigrant families already resident in the state. He even offered to provide further federal assistance to mitigate the cost to the state of new immigration, but the so-called 'Save our State' initiative was approved in November 1994. The new measure, introduced when the devaluation of the Mexican peso was expected to increase migration, would force social benefits, including education, to be withdrawn from illegal immigrant families in an attempt to deter migration.

Not surprisingly, the migration issue soon disrupted US–Mexican relations as the number of people denied entry into the US from Mexico at border posts rose by 26% between October 1994 and June 1995. In July 1995 another irritant emerged. Although Congress did not achieve the two-thirds majority needed to override a presidential veto, it voted by 245 to 183 to prohibit the Treasury Department from extending any further assistance to Mexico from its Exchange Stabilization Fund.

Similarly, the exacerbation of the drug problem in Mexico has not only confirmed the salience of the issue for US–Mexican relations, but has become a useful weapon for Clinton's critics. Mexico is now the main route for cocaine entering the US; 70–80% now comes from Mexico. The apparent links that have been uncovered between drug-traffickers and both the police corps and members of the armed forces, the increased violence – including high-level assassinations – accompanying Mexico's efforts to combat drug trafficking, and more recent allegations of links between senior politicians and the drug cartels, have all cast a pall over US–Mexican relations.

Under the rescue package umbrella, the Mexican government has agreed to increase its cooperation in the war against drugs. Thus far the Clinton administration has relied on collaboration at various levels, but pressure in favour of a firmer stand has been mounting. This has been particularly clear in the initiatives submitted to the US Congress linking financial assistance to progress in the war against drugs. Despite Clinton's attempts to establish a sound basis for managing the bilateral relationship and for keeping Mexico from becoming an election issue, the charged political climate in the US throughout the electoral year will make it difficult to achieve these goals.

A Possible Brighter Political Future

Despite the eruption of the Zapatista rebellion into Salinas' strategy of ordered gradual change, and the dramatic symptoms of instability observed throughout 1995, the growing institutionalisation of political competition over the past two years could become an important source for order and stability for Mexico. The consensus that the 1994 elections were relatively clean increased the pressure on political parties to operate within the institutional framework. In January 1995, the PRI, PAN and PRD signed the Commitment for a National Political Accord to promote balanced relations between the executive, legislative and judicial branches of government, to negotiate a 'definite electoral reform' and to establish an impartial electoral authority.

President Zedillo used the occasion to criticise Salinas' management of political change, and then radically departed from the practice of previous PRI governments by appointing a PAN member as General Attorney. Although implementation of the Commitment was delayed by post-electoral disputes in Tabasco and Yucatán, the process was re-established in October 1995. By November, the parties had agreed on a general framework to discuss political reform, with negotiations focusing on the autonomous electoral authority and a new electoral law to regulate the 1997 mid-term elections.

Mexico's opposition parties are now more robust. Competitive elections are increasingly the norm rather than the exception, and post-electoral conflicts – a mechanism often used during Salinas' government to resolve electoral outcomes – have not occurred since the 1994 presidential elections. Electoral rules, traditionally the object of endless legal debate, are now decided through suffrage.

A future victory at the polls by the opposition cannot be ruled out, especially if the economic stabilisation policies fail to deliver and there is no growth by 1997. It is also likely, however, that this institutionalisation of political competition may act as a safety valve for some of the disruptions and turbulence that have recently jolted the political system. Ironically, Salinas' decision to open up the political system gradually may provide Zedillo with an instrument and the time needed to channel the social and political discontent which will continue to accompany economic adjustment into a fuller acceptance of the reforms. His capacity to capitalise on this option will largely depend on his ability simultaneously to contain the disarray within the PRI and internal strife within the country.

Europe

Any hopes that Europe would emerge as an area of renewed stability were thwarted in 1995. By early 1996, the war in Bosnia was far from resolved; the European Union remained divided over key issues of policy; and prospects for continued reform in Russia had dimmed.

NATO appeared both to have surmounted the difficulties it had faced earlier in the year and to be playing the lead role in the quest for lasting peace in the Balkans through the deployment of its Implementation Force (IFOR) to Bosnia at the end of 1995. This was the only institutional success story, however. The European Union approached its March 1996 Inter-Governmental Conference (IGC) divided over issues of European Monetary Union (EMU), a Common Foreign and Security Policy (CFSP), eastward enlargement and institutional reform. The Western European Union (WEU), despite the appointment of a new Secretary-General and its continued role in policing the Adriatic, remained marginalised and over-shadowed by NATO's greater contribution to European security, as well as by other bilateral and multilateral initiatives between individual EU member-states, such as the creation of the Franco-British Rapid Reaction Force for Bosnia.

In Russia, the most worrying trend of 1995 was the return to prominence of the Communist Party, led by Gennady Zyuganov. President Boris Yeltsin's grip on power is no longer assured. It has become clear that his health is in decline, and he no longer commands the authority at home that he once did. Yet he cannot be counted out. A new factor in Russia's domestic politics is the rise of economic interests, with power and influence going to those regions and companies that have acquired new prosperity and wealth under Yeltsin's reforms. Although still beset by problems, the Russian economy showed promising signs of stabilisation, and how it develops in the first half of 1996 will play a significant part in the presidential elections in June.

In the Balkans, 1995 began with no sign of an end to the conflict that had raged since 1991, but ended with the signing of a peace agreement on 14 December in Paris. This, however, does not signify a return to normality. IFOR may well preserve the fragile peace in Bosnia until the end of its mandate in November 1996. But without a greater commitment and intervention from the international community, the signs for lasting peace in the region are by no means encouraging.

Western Europe's Search For The Elusive Grail

The gap between the rhetoric and the reality of European integration widened in 1995. The European Union was supposed to be laying the foundations for its 1996 Inter-Governmental Conference, which began on 29 March in Turin, to streamline decision-making and strengthen the EU's Common Foreign and Security Policy. Instead, the Maastricht Treaty timetable for monetary union began to unravel and a fledgling European security policy was eclipsed by the US role in ending the war in Bosnia.

The Maastricht timetable for creating a common currency ran up against economic reality in 1995. Not only did it appear that Luxembourg alone would join Germany in meeting Maastricht's stringent fiscal criteria by 1997, the earliest starting date foreseen by the Treaty, but it became increasingly unlikely that more than a handful of countries would meet the criteria by the final deadline of 1999. Reducing annual public deficits to under 3% of gross domestic product (GDP) and total public debt to under 60% of GDP, as called for in the Treaty, was proving extremely difficult in a Europe plagued by bloated public sectors and high unemployment. Budget cutting led to massive strikes in France, and in Germany, widespread public opposition to abandoning the Deutschmark began to be echoed by the country's political leaders.

Four years of failure to achieve a negotiated peace in Bosnia and the enduring difficulty of finding common ground among the EU's 15 members cast a shadow over hopes of achieving a CFSP with real influence. Aspirations for greater European independence in the security sphere became little more than a desire to give NATO a more European flavour. Even France became more Atlanticist. With the EU's two major projects – European Monetary Union and the CFSP – on the rocks, momentum on the eve of the IGC was slow.

National Difficulties

European Union economies grew at a healthy 2.7% in 1995, but two consecutive years of strong growth could only make a small dent in stubbornly high unemployment rates, which averaged 10.7%. Nor did economic growth buoy enthusiasm for 'ever closer union'. Almost every EU member-state found that tackling unemployment while aggressively pursuing the budget cuts and low inflation demanded by the Maastricht criteria for monetary union was not feasible. Whatever grand plans and high hopes had existed before the 1996 IGC were badly deflated.

The driving force behind a federal European Union bound together by a common currency was clearly Chancellor Helmut Kohl's Germany. Nevertheless, the country was still struggling to recover its identity as the fiftieth anniversary of Hitler's defeat was celebrated throughout Europe. Although during 1995 Germany was constantly reminded of a past it would like to exorcise, it did its best to shift attention away from the past and towards Europe's future. Bonn's foreign policy was informed by a vision of European Union, even if it had considerable difficulty formulating a strategy to contend with the dilemmas inherent in that vision. Another difficulty was how Europe might evolve if monetary union were not possible, and if other nations were less inclined to create a 'Federal Republic of Europe'. Germany's influence might have been increasingly felt, but a coherent pattern of leadership had yet to emerge.

Helmut Kohl enjoyed sustained public support in 1995, due not so much to his own dynamism as to the opposition Social Democratic Party (SPD's) enduring internecine struggles. Kohl's domestic popularity was so high that, in the three regional elections held at the end of March – which his party won with ease – he was able to help the Free Democratic Party (FDP), his coalition partner in parliament, to a good showing after an early projection had suggested they might suffer another electoral disaster. The Green Party had replaced the FDP as Germany's third strongest political force, partly by introducing relative moderation into its stand on security and defence. In December 1995, half the Green *Bundestag* deputies followed floor leader Joschka Fischer by voting in favour of Germany's 4,000-troop contribution to NATO's peace implementation force, IFOR, which was to deploy to the former Yugoslavia.

While the Greens grew more moderate, the SPD embarked on a more radical course when Oskar Lafontaine staged a dramatic coup at the November party conference and replaced Rudolf Scharping as party leader. One clear sign of this shift in policy was Lafontaine's opposition to German participation in IFOR (or in any other out-of-area) combat operations. In the December *Bundestag* vote, one-third of SPD deputies backed him in opposing German participation. Over half of these deputies had already opposed Kohl's June 1995 decision to provide electronic combat and reconnaissance (ECR) *Tornados* to support the UN Protection Force (UN-PROFOR) and the Franco-British Rapid Reaction Force in Bosnia. The deployment of German forces was the country's first out-of-area combat role and, notably, occurred in the former Yugoslavia where Kohl had long said German troops would never serve.

In autumn 1995, Gerhard Schröder, SPD Minister President of Lower Saxony, urged a delay in European monetary union, thereby spearheading a move within the SPD to capitalise on widespread public opposition to abandoning the Deutschmark. An opinion poll taken in late 1995 showed that opposition to the new currency, the 'euro', had climbed to 85%. The SPD's objection to the use of German combat troops abroad and the temptation in both the SPD and Bavaria's Christian Social Union (CSU) to turn the public resistance to monetary union into a 'national' electoral issue, thus challenged Kohl on two fundamental pillars of his European policy.

In 1995, Jacques Chirac's campaign for President of France and his forceful actions since his election in May dominated French domestic politics. By revamping French security policy he has realigned European geopolitics. When he moved France a great step closer to NATO, at the same time announcing a leaner, more mobile and all-volunteer military, Chirac – in contrast to his predecessor – demonstrated more interest in increasing France's influence in the Atlantic Alliance than in building an independent European defence capability.

In the period leading up to the 23 April and 7 May presidential elections, Chirac promised to reduce unemployment, heal the 'social fracture' and raise public-sector wages while cutting the deficit and lowering taxes. These incompatible objectives led to a 'zigzag' policy. First, Alain Juppé's government called for gradual reform, sacking Finance Minister Alain Madelin for pushing austerity too strongly. Then, at the end of October, Juppé shifted his priorities, presenting a budget that delayed tax cuts and reduced public-sector wages and pensions. His intention was to reduce France's annual public deficit from over 5% to under 3% to meet the Maastricht criteria for monetary union. Instead, his action caused the greatest wave of unrest since 1968 as public-sector workers mounted massive strikes against the cuts. The strikes finally subsided, even though the government's few adjustments left the austerity plan more or less intact, but the experience underlined the difficulty France would have in maintaining social peace as it sought to meet the stringent fiscal conditions of the EMU.

Chirac entered office determined to reassert French power and influence. His decision to carry out a series of nuclear tests in the South Pacific, however, damaged France's prestige, particularly in Asia. Chirac's assertiveness on Bosnia was of greater long-term consequence.

Berating the United States, Europe and the United Nations for their failure to improve the situation in Bosnia, Chirac threatened to pull French peacekeepers out of UNPROFOR if they were not given the means to protect themselves and carry out their mandate. Throughout summer 1995, particularly after the fall of Srebrenica to the Bosnian Serbs, France played an instrumental role in pressing for a more assertive policy on Bosnia. Ironically, France helped set in motion a chain of events that ultimately gave NATO a new mission and sense of purpose in Bosnia. France's decision to rejoin the Alliance's integrated military command after 30 years of absence indicated, more formally, its acceptance that NATO could only be influenced from inside.

While Chirac's strong stand on Bosnia earned him respect at home and abroad, his nuclear tests did not. Though he had not consulted his partners before announcing the test series in June, he complained loudly when ten of France's EU partners voted in favour of a November 1995 UN resolution condemning the tests. France's attempt to smooth matters over in Europe by introducing the concept of 'concerted deterrence' which could, in time, lead Germany to contribute to nuclear policy-making in France, was examined with extreme caution in Germany. Nevertheless, this offer, and France's growing willingness to cooperate more closely with the UK on nuclear issues, demonstrated the increasingly important role that Europe had come to play in France's nominally independent nuclear policy.

In another effort to adjust to new realities, President Chirac announced in February 1996 plans for a dramatic reform of France's armed services. Recognising that power projection could only be assured by creating a truly professional army, Chirac set in motion a 'revolution' in French defence policy. Conscription will be abandoned over a five-year period, the armed forces will shrink by one-third – from 400,000 to 250,000 – and a 50,000-strong crisis-reaction force will be created. The land-based leg of the nuclear deterrent will be dismantled, and France's defence industry will be consolidated: Aerospatiale and Dassault will merge and the Thomson electronics group will be privatised. The creation of a truly professional army – 'on the British model' – will in the long run make France more self-confident in calling for a strong European pillar to the Atlantic Alliance.

In the UK, Prime Minister John Major's slide to political disfavour continued. He survived a Conservative Party vote to replace him as leader in July 1995, but by February 1996 his party's majority in parliament had dwindled to two, leaving him all but paralysed on issues of any controversy. This was particularly the case over Europe, where Tory sceptics blocked any UK move into the European Union mainstream, particularly on monetary union. The UK White Paper on the IGC, issued on 13 March 1996, demonstrated that the government now took a steady Eurosceptic line.

On Ireland, too, Major had his hands tied. Throughout 1995, he linked the possibility of all-party negotiations to an Irish Republican Army (IRA) agreement to disarm. In January 1996, an international commission under former US Senator George Mitchell recommended that disarmament and talks should proceed simultaneously. Major side-stepped this recommendation by suggesting that elections of an undetermined nature be the next step in Northern Ireland. The IRA, which condemned the election concept as a plan originally proposed by the Ulster Unionists, reacted by breaking the cease-fire with a series of bombings in London in February 1996. The UK and Irish governments attempted to reorganise all-party talks, scheduling them to begin on 10 June 1996 and even allowing Sinn Fein to attend if the IRA would renounce violence. At the end of March, the terrorists reiterated that they were not prepared to give up violence and join the peace process.

Although UK elections do not have to be held until March 1997 and much can change before then, the signs were that the electorate felt the government had been in power for too long. Opinion polls taken in spring 1996 showed the Conservative Party trailing the rival Labour Party by at least 25 points. Prime Minister Major's personal support was no higher than 30%, while Labour leader Tony Blair enjoyed a surge of popularity with his push for moderation and modernisation. On Europe, Blair distinguished himself from Major by his greater receptivity to UK participation in the EMU and by his desire for UK participation in the social chapter. Otherwise, Blair's cautious stance showed sensitivity to popular scepticism about the EU and hinted at enduring differences within his own party.

A distinct Mediterranean emphasis characterised the European Union in 1995 and early 1996 because of the influence of the French, Spanish and Italian EU presidencies. The countries of the European Community's Mediterranean expansion in the 1980s – Spain, Portugal and Greece – had consolidated democracy at home and established these countries' place in Brussels, but in the first months of 1996 they were again at a difficult juncture. The long socialist reigns of Gonzales in Spain, Soares in Portugal and Papandreou in Greece came to an end. Moreover, these countries had to confront the prospect of exclusion from 'Europe's' common currency.

So did Italy, despite the initial success of Lamberto Dini's reform government in cutting public spending and the pension system. By the beginning of 1996, Dini's efforts had run out of steam, and he stepped down on 11 January. With Italy's election system still unreformed, political fragmentation and instability promised to accompany Italy throughout its EU presidency. The country's domestic distractions thus placed one more obstacle before the EU's Inter-Governmental Conference.

The European Union

Debate over the European Union's future dominated its political agenda and the 1995 European Council summits in Cannes and Madrid. A flurry of studies from the Commission, the Parliament and the Council's Reflection Group attempted to pave the way for the difficult decisions on monetary union, expansion and institutional reform that lay ahead. One of the difficulties was clear. At their meeting in Cannes at the end of June, EU leaders acknowledged that the majority of countries would not fulfil the Maastricht criteria for EMU by 1997. Those countries that did, according to the Maastricht Treaty, would automatically enter monetary union in 1999.

The Cannes meeting had been overshadowed by the results of the French presidential elections; the 15–16 December summit in Madrid was more ambitious. The EU finance ministers and heads of state reaffirmed their commitment to a common currency by 1999, officially named it the 'euro', and laid out a detailed timetable for its introduction. While they agreed to make decisions on who would join 'early in 1998', they said little about the growing doubt that any member except Germany and Luxembourg would fulfil the criteria by then. In another ambitious Madrid commitment, EU leaders joined nine Central and East European heads of state to declare the EU's 'hope' of starting membership negotiations six months after the end of the IGC.

In 1995, the European Union – under the French and Spanish presidencies – shifted its focus from Eastern Europe to the southern Mediterranean, where Islamic fundamentalism combined with underdevelopment threatened the prospect of mass migration or worse. Algeria's continuing civil war underlined this threat all too clearly. On 27–28 November, EU foreign ministers met in Barcelona with their colleagues from Morocco, Tunisia, Algeria, Egypt, Israel, Jordan, Lebanon, Syria, Turkey, Malta and Cyprus, as well as Palestine Liberation Organisation (PLO) Chairman Yasser Arafat. They committed themselves to more multilateral contacts, a significant increase in aid (to 2000, EU aid to the Mediterranean will be 4.85 billion ECU; by contrast, 6.69bn will go to Central and Eastern Europe), and a modest agreement on free trade by 2010 (omitting agricultural goods). Yet, the Barcelona conference once again demonstrated the EU's difficulty in complementing development aid with free trade. Without access to EU markets, particularly agricultural markets, economic development of its eastern and southern periphery would remain slow at best.

Trade and aid policy characterised the European Union's interaction with the outside world, but security issues, whether domestic or international, found less of a place in EU policy. Efforts to establish a coordinated European police force moved slowly, a common immigration policy was still a distant desire, and security and defence issues were largely relegated to NATO.

Ad hoc initiatives dominated European defence cooperation in 1995. The Western European Union acquired a new member, Greece, and a new Secretary-General, Portugal's José Cutileiro. The organisation also continued its operational role in jointly enforcing the naval blockade in the Adriatic and helping to police the embargo on the Danube. There was no mistaking the fact, however, that the WEU was still only on the periphery of Europe's major military efforts, and many, even in France, began to doubt its utility.

European military activity remained for 'the coalition of the ready and the willing'. The UK and France cooperated in creating a Rapid Reaction

Force for Bosnia and agreed to closer consultation on nuclear issues at a 30–31 October summit between John Major and Jacques Chirac. France and Germany also continued to expand their security partnership. They agreed to establish a joint armaments agency in Bonn, thereby all but abandoning the WEU's floundering efforts to set up a larger European Armaments Agency. They also decided to cooperate in developing two reconnaissance satellite programmes.

The Eurocorps they initiated became operational on 30 November 1995 and carried out its first exercise with troops from France, Germany, Spain and Belgium and with observers from the new member, Luxembourg. Building on the Eurocorps experience and reflecting Europe's Mediterranean focus, France, Italy, Spain and Portugal agreed to set up two new forces: EUROFOR, a division of 10,000–15,000 troops; and EUROMARFOR, a non-permanent air and naval task force. But the Eurocorps' future was later called into question by France's decision to reduce and professionalise its force structure.

Facing Tough Choices

In the year leading up to its March 1996 Inter-Governmental Conference, the EU was confounded by three difficult issues: European monetary union; enlargement to the east; and institutional reform.

With most EU economies still far from meeting the Maastricht criteria for monetary union, the likelihood increased that either the 1999 deadline or the criteria for membership would have to be changed. Even then, with the Kohl government's strong desire for a single currency matched by its firm insistence that the German model of fiscal and monetary discipline should underpin that currency, Europe seemed to be heading for a monetary union confined to Germany and only a small number of other EU members. France's efforts to adhere to the stringent German criteria fuelled domestic unrest and threatened the Franco-German axis. The political imperative of including France in a monetary union clashed directly with the increasingly strict economic conditions Germany (particularly in the figure of Finance Minister Theo Waigel) demanded of such a union.

The politics and the economics of EU expansion were also in conflict. The December 1995 Madrid conference committed the EU to begin negotiations six months after the end of the IGC; in July, Kohl had promised Poland (and, by implication, other Central and East European countries) to negotiate an agreement on membership by 2000. At the same time, a number of European Commission studies underlined the problems expansion would pose, particularly for the EU's budget, but also for its decision-making mechanism and its fledgling security policy.

The EU's expansion from 12 to 15 members in 1995 to perhaps 20 or more thereafter, and the ensuing need to streamline decision-making, was one of the main reasons for scheduling another IGC. The 1996 Inter-Governmental Conference will attempt to increase the efficiency of EU decision-making by rebalancing the power of the Council, Commission and Parliament. There were almost as many definitions of greater efficiency in 1995, however, as there were Union members. The Reflection Group, headed by Spain's Carlos Westendrop, outlined a series of possible reforms, particularly increased majority voting. Released prior to the Madrid summit, the report – which only outlined the issues and did not make recommendations – was more a portrayal of the differences among EU members than a guide for overcoming the Union's major dilemmas.

There was a fundamental divide between those countries that sought a federal Europe, such as Germany, and those, especially the UK, that sought a Europe of closely cooperating, but independent, nation-states. France found itself in the middle, wooed by both the UK and Germany. The close cooperation between Germany and France that had preceded the Maastricht Treaty was not so apparent in the months before the IGC.

The two countries did, however, issue a memorandum in early December 1995, and again in February 1996, declaring their common goals, including strengthening EU's foreign and defence policy. Yet the vagueness of these memoranda clearly betrayed the divergence of views within the partnership. Although Germany and France both wanted a stronger Union, Germany sought a more influential parliament and more majority voting, while France was reluctant to support majority voting and wanted the EU's ultimate power to remain in the Council.

Another divide separated Europe's large and small countries over representation and voting power. Growing membership made it increasingly necessary to prevent the veto of a single member holding back the EU, as Greece's position on the former Yugoslav Republic of Macedonia had so clearly shown. Smaller countries were reluctant to be excluded from the Commission, with only a small voice in the Council.

Some of the greatest differences surrounded the question of foreign and security policy. The more federally oriented member-states wanted to bring it, as well as justice and interior issues, under the Community's supranational umbrella. On EU foreign policy, Germany, Italy, Spain and most smaller countries wanted non-defence decisions taken by majority vote. The UK and France did not. In turn, the Commission wanted to create a larger role for itself, a highly unlikely prospect as the UK and France strove to keep foreign and security policy strictly inter-governmental. The UK and France, however, remained divided on whether the WEU should be integrated into the EU (France) or kept separate (the UK). Broader consensus did exist on the necessary, though hardly sufficient, step of setting up a Brussels-based security-policy planning staff for the European Union, and the possible creation of a spokesperson on European foreign policy.

What became clear in 1995 was that with such a wide variety of views on reforming the European Union, reaching a consensus for substantial reforms at the 1996 IGC would be very difficult. At the same time, it became increasingly clear that Germany intended to play a larger role in the forthcoming IGC than it had at Maastricht. By contrast, the Franco-German tandem appeared less influential.

Finding a Focus

NATO, too, faced a series of uncertainties, although it seemed to cope with them better than the EU. This was partly because it had found a new sense of purpose in its role in Bosnia. The Alliance had entered 1995 still reeling from the division provoked by the United States' decision to stop enforcing the arms embargo against the former Yugoslavia. US threats to lift the embargo unilaterally and apparent European ineptitude in dealing with the Bosnian war continued to place a severe strain on transatlantic relations. This was exacerbated by the contentious issues of NATO expansion and Europe's security role. Some of this tension was alleviated when the US brokered an end to the fighting in Bosnia and NATO sent in the powerful IFOR to keep the peace that had been achieved.

In many ways, it was France that broke the transatlantic log jam on Bosnia. France's push for a stronger Western response and the Clinton administration's growing desire to see the Bosnian war end before the November 1996 US elections moved the Americans once again to focus on Bosnia. The fall of Srebrenica and the ensuing mass killing of some 5,000 Bosnian males had also brought about a change of heart in the UK. The US, France and the UK thus agreed to move beyond pin-prick attacks and to threaten 'disproportionate' strikes against a wide range of Bosnian Serb targets should they continue to violate the remaining safe havens of Gorazde, Sarajevo, Tuzla and Bihac. NATO ministers endorsed this new position at a meeting in London on 21 July.

NATO was thus prepared when the Bosnian Serbs fired a mortar round into a Sarajevo market-place on 28 August leaving 37 dead. On 30 August NATO launched *Operation Deliberate Force*, a two-week bombing campaign and artillery attack from the Franco-British base on Mount Igman officially aimed at forcing Bosnian Serb Commander General Ratko Mladic to move his heavy guns away from Sarajevo. Flying more than 3,000 sorties against Bosnian Serb air defence, command and control, and ammunition dumps, NATO aircraft also badly battered Mladic's army, leaving it vulnerable to a Muslim–Croat offensive in north-west Bosnia. By the time of the final cease-fire agreement on 12 October, this offensive had reduced Serb-held territory in Bosnia from 70% to less than 50%. While US Assistant Secretary of State Richard Holbrooke's shuttle diplomacy played a key role in cementing these agreements, the air and ground campaigns demonstrated that the situation on the battlefield does indeed greatly influence the outcome at the negotiation table.

The Dayton accords, initialled on 21 November in Dayton, Ohio, pointed the way towards peace, but did not guarantee it. NATO, having played a major role in bringing the warring parties to agreement, now played an even greater role in ensuring they adhered to it. Following the 14 December peace treaty signed in Paris, NATO began to move 60,000 troops from 32 countries into Bosnia to police a 1,100km zone of separation. Despite difficult weather, *Operation Joint Endeavour*'s build-up in Bosnia went well. It met no armed resistance and by March 1996 had managed to carry out all the tasks assigned for that period.

NATO's initial success in Bosnia certainly gave the Alliance a sense of accomplishment and purpose. Still, a number of difficult issues continued to feed NATO's underlying post-Cold War uncertainty. Transatlantic sensitivities remained just below the surface. They had flared earlier in the aftermath of the resignation of NATO General Secretary Willy Claes on 20 October. When the US vetoed the 'European's' candidate, it took until 5 December for NATO to agree that Spanish Foreign Minister Javier Solana should fill the post.

France's move towards NATO in 1995 marked a turning-point in the policy of independence established by General de Gaulle 30 years earlier. First, France placed 7,000 of its troops under NATO's IFOR command, and then on 5 December declared that it would henceforth participate fully in NATO's Military Committee and work more closely with Supreme Headquarters Allied Powers Europe (SHAPE) – ostensibly to further NATO reform.

At the same time, France remained at odds with other Alliance members, particularly the United States, on the make-up of Combined Joint Task Forces (CJTFs). France has long insisted that NATO assets should be part of CJTFs, even if non-NATO headquarters, such as the Eurocorps, were used. The US, however, does not want to provide assets to headquarters it does not control.

NATO also remained entangled in the debate over enlargement. The Partnership for Peace (PFP) programme was proving an effective means for enhancing military cooperation and not just a way of holding off new members. Nevertheless, the issue of enlargement remained an irritant both within NATO and to the Russians. The publication in September of the 'Study on NATO Enlargement' did little more than reaffirm the Alliance's position that enlargement would occur. It focused on the 'how' and 'why', but said nothing about the 'who' and 'when'.

Even so, it met with strong Russian opposition – particularly since it came on the heels of the bombing of Serb positions in Bosnia. Nevertheless, NATO continued to seek a special relationship with Russia as a partial palliative. Russian participation in IFOR offered the hope that substantive cooperation could remove some of the excessive rhetoric.

Outlook

The closer the European Union came to its March 1996 Inter-Governmental Conference, the more apparent it became that the Union faces a number of daunting dilemmas: creating a monetary union that works but does not divide; enlarging the EU without undermining its effectiveness; and increasing decision-making efficiency without hollowing out national sovereignty. Europe's stubbornly high unemployment rates sharpened the realisation of just how difficult it was going to be to resolve these dilemmas. The recognition that it would be a long and hard road slowed much of the momentum in the run-up to the long-heralded IGC.

Because of the vexing issues it faces, the Conference may last even longer than the expected 12–15 months. Some participants may even be willing to see it proceed slowly in the hope that the anticipated UK general election will produce a more pro-European government. Despite Germany's intentions and the efforts it is expected to make, the IGC is unlikely to take any significant step towards supranational federalism. More majority voting may be added here and there, the European Parliament will probably acquire some new, but still limited, rights, and while foreign and security policy might gain a modicum of institutional foundation, it will do so without really losing its inter-governmental essence.

Changes of this kind will help streamline decision-making for a Europe Union of 15 members, but will by no means create a structure capable of absorbing 5–10 new members over the next decade. Meanwhile, France and Germany's current insistence that European Monetary Union should be initiated in 1999 will, at a minimum, divert an increasing amount of political attention from other pressing European issues. And the chances that the Franco-German position will be successful is very small indeed. Delaying the 1999 deadline is far more likely.

A real Common Foreign and Security Policy for the European Union also looked increasingly unlikely in 1995, not only because of the EU's own troubles, but because of the United States' decisive role in ending the Bosnian war. Where Europe did make progress on defence, it was among smaller groups of states cooperating on an *ad hoc* basis. The *loci* of security cooperation were either France and Germany, France and the UK, or the Mediterranean countries. It was not the WEU.

NATO was the one institution in Europe whose leaders could face 1996 with satisfaction. NATO had entered 1995 deeply divided over the Bosnian war, but an unusual interplay between France and the United States had put it back at the centre of European security. The Alliance had made little progress in resolving the issue of expansion. Yet the success of the Bosnian operation did replenish the political capital NATO will need to contend with expansion. Of equal importance, IFOR was helping to forge substantive bonds with East European countries – including Russia – through concrete cooperation on the ground. Unlike the other European institutions, NATO at last has some reason for self-confidence.

Troubled Times For Russia

A year that began with the invasion of Chechnya and ended with the communists dominating the Duma cannot be considered a good year for Russian reform. In 1995, all the contradictions and dangers inherent in the country's attempted transformation were revealed: an economy that made significant progress towards stabilisation, but still lacked basic legal and administrative structures; an enfeebled central state led by an increasingly enfeebled president; a yawning social gap between the haves and have-nots; the continued growth of nationalist and communist political forces; the omnipresence of crime and corruption; and an assertive foreign policy that focused increasingly on restoring Russia as a great power.

These trends have led many to conclude that Russia is on the verge of a great reversal. Yet the market has begun to plant its roots in Russian life. Political power is no longer a state monopoly, but has itself fragmented within the government, between the centre and the regions, among new economic and political interests, and into a fledgling civil society. This society is active and vibrant, even if its influence over the day-to-day actions of the government is limited. These factors, as well as profound limitations on the political, economic and military capabilities at hand, make a great reversal unlikely. Russia, instead, is likely to remain enmeshed in a long and uncertain transition.

Worrying Indicators

1995 ended with a bang. Over 60% of Russia's eligible voters turned out in mid-December for parliamentary elections that gave the Communist Party the largest single share of the votes (over 20%) on the party lists and significant success in single-mandate districts. Only one of the two government parties that had been created with such fanfare in the spring, Prime Minister Viktor Chernomyrdin's 'Our Home is Russia', made it over the 5% threshold. Over one-third of the 450 members of the new parliament, half elected from the party list and half elected in single-mandate districts, are communists or belong to parties politically in tune with them.

Given this expression of the public will, Communist Party leader Gennady Zyuganov is in a strong position to emerge as one of the two top candidates in the first round of the June 1996 presidential elections, and thus is a strong contender to become the next president of Russia. Following the December 1995 elections, current President Boris Yeltsin clearly noted the electorate's warning and embarked on a strategy that mixes 'populism, great-power themes and quasi-liberalism' designed to seduce the electorate. He has rid himself of the liberals Andrei Kozyrev and Anatoly Chubais, the two men most associated with reform. He has promised to increase old-age pensions and to pay wage arrears to government and industrial employees. But even with these policy shifts, the election is not Yeltsin's to lose. He faces competition in the first round of voting not only from Zyuganov, but also from General Aleksandr Lebed, Vladimir Zhirinovsky and Grigoriy Yavlinskiy. In polls taken in the first months of 1996, no candidate except Zyuganov registered above single figures.

Yet 1995 was more than a prologue to the 1996 presidential elections. Developments throughout the year illustrated the basic trends that will continue to shape Russian politics in the years ahead. Russian politics is like a three-layered pyramid. At the top is Yeltsin and his entourage, the key political figures shaping state policy and the state itself. In the middle is a mixture of old and new institutions, interests and players that compete for the state's favour, for its property and for control over its policy. At the base of the pyramid are the people.

The top two layers form the bulk of the news, but the very fact of elections, even as an obstacle to be endured, fought against or, in the view of some, cancelled, is a new aspect of Russian political life. So is the interaction between and within these three layers.

At the top, the question of Yeltsin's physical and political health has been a constant theme of speculation. Yeltsin was hospitalised for minor nose surgery, and thus was absent from public view, during the opening days of the Chechen invasion in December 1994. When he did emerge from hospital, he appeared to be in shaky command of both the facts and his own troops, claiming to have halted the bombing, even as the world tracked Russian air assaults on television. An unsteady and faltering Yeltsin appeared at a February 1995 Commonwealth of Independent States summit in Almaty, Kazakhstan. Reports that his appearance did not betoken illness so much as heavy drinking on the plane from Moscow did little to calm concern for Yeltsin's grip on power. From conducting the band at a commemorative ceremony in May in Berlin to an absurdly jovial press conference with US President Bill Clinton in New York in October, Yeltsin regularly appeared either out of touch or out of control.

Although at other times Yeltsin showed himself firmly at the helm, two serious heart incidents in the summer and late autumn removed him from the political scene for long periods, including during much of the parliamentary election campaign. Although he later announced that he was running for re-election, his health remains the real wild card. If he were forced to drop out, it would truly reshape the campaign in unpredictable ways.

Whether for health or other reasons, Yeltsin increasingly let his hands slip from the tiller during 1995. A coterie of advisers stepped in to fill the void, including Presidential Assistant Viktor Ilyushin, an old Yeltsin aide from Sverdlovsk, Generals Alexandr Korzhakov and Mikhail Barsukov, who have leap-frogged from bodyguards to positions of extraordinary influence over policy, and First Deputy Prime Minister Oleg Soskovets.

These men have been increasingly visible in public policy, with Korzhakov serving for a time, along with Yeltsin's wife, as the public go-between during the President's seclusion in hospital. These advisers have emerged from the old power centres of the Soviet state, such as the KGB, the Communist Party or the military–industrial complex and in 1995 they forged important new links with segments of the business and financial community. Though many have sought to describe the rise of these men as the victory of an anti-reform or war party, their presence has not rejuvenated the state or the institutions they are really supposed to represent.

In fact, the state remains immeasurably weak. In part, the flow of power away from the centre is a sign of reform. The state in Russia, as one historian has noted, always 'waxed fat, while the people grew lean'. This is in part the natural result of the old power structures weakening before their successor organisations are firmly in place. But the current political struggles at the top and middle for property, privilege and position continue to weaken the state. The state remains an arbiter of key policy matters, but is no longer what the Russians refer to as the *khozyain* (master) of them. It must negotiate with the regions over fundamental powers and revenues; it must mediate between increasingly independent ministers over dwindling resources; it can cajole and threaten – but it cannot always command.

What is new is the rise of different economic interests. Real wealth has been created in Russia. The banks are a major factor of contemporary life, as are independent entrepreneurs. The gas and oil lobbies have their own corporate interests in Caspian oil or tax policy that do not necessarily coincide with the government's policies. *Lukoil* still figures in the coalition that decided in October 1995 to split early Caspian oil between the Russian pipeline and a new route through Georgia and Turkey, a position the Russian government opposed. Rich regions have also emerged with real influence because of the natural resources they control. But even poorer agricultural regions have obtained subsidies and other favours from the centre, demonstrating that even the enfeebled have some leverage.

A few key 'clans' or 'cartels', such as the oil and gas lobbies, the major banks, and strong regions – for example Moscow or Tatarstan – are at the centre of Russian politics. These large interests have the wealth and access to win a predominant share of the country's spoils. While they play a key role in building political coalitions inside Moscow, they are far from a permanent plutocracy. Instead they are themselves a disparate group united by greed, and their position in Russian political life is nowhere near as imposing as the Party and core institutions that ran the Soviet state.

These groups illustrate that the struggle at the heart of Russian politics is not between reformers and anti-reformers. Instead, it is between groups grasping for a share of the economic wealth and political power in the new Russia by intervening within a weak regulatory environment and a still developing market framework vulnerable to rent-seeking and monopolistic arrangements. So far, they have tended to intervene not by long-term investment and product development, but by controlling the privatisation of the state's assets.

One observer has called this process 'the struggle of rent-seekers and reformers'. The activity of the powerful interests and their government patrons weakens attempts to consolidate the Russian economy and government by eroding the basis for fair competition. The interest groups oppose new regulations, taxes and even the introduction of full market conditions that would erode the privileges they have already gained. Their dominance in key and high-profile economic questions also further exacerbates the gap between rulers and ruled. The rich are becoming immeasurably richer, while the poor fear that the modest social services provided by the old regime are already crumbling.

The people at the bottom of the pyramid enjoy great freedom from mass repression and forced political behaviour compared to the past. They have been emancipated from state control, but still have little real influence over the day-to-day decisions of government at either the national or regional level. But they control one significant lever in the process – elections – and they spoke loudly in December.

The current government knows it is genuinely at risk in the forthcoming presidential race. The cancellation of the election altogether, or the manipulation of its results, cannot be ruled out. The benefits enjoyed by *Gazprom* or the major banks could also be undone or modified by a change in leadership. But it would be difficult to postpone or cancel the elections without raising profound questions of legitimacy and control. Elections have become, or nearly become, an unquestioned part of the political process in Russia.

Thus, in the June 1996 elections Yeltsin and his rivals will see tested the system that they and the new political actors have helped to create. It remains at best a mixed system, with important democratic elements of a significance found in no stable democracy precisely because they are un-supported by the normal constraints of a fixed and tested system of laws and of a state that is strong, not in the traditional Russian sense, but as an entity that can legislate and enforce the rules of the game.

Struggling with the Economy

The Russian economy made considerable progress towards stabilisation in 1995. GDP declined by a further 4% from its 12.6% fall in 1994. Some analysts believe that this 4% decline represents the beginning of a real, if hidden, growth that official statistics cannot measure. Inflation, which had been running at 17% in January 1995, dropped to 3.2% in December. The annual inflation rate of 130% was a distinct improvement on the 215% rate for 1994 and 840% in 1993. Russia also had a large trade surplus of over $20 billion, mainly as a result of exporting raw materials. The November 1995 figures for foreign investment show a 112% increase over the same period in 1994, although the problem of capital flight remains. Basic economic progress convinced London and Paris Club creditors to reschedule over $70bn in Russian debt in November 1995.

That was the good news; but there was plenty of bad news. Real income fell by 13% in 1995 compared with 1994. Official figures registered 2.3 million people – or 3.1% of the working population – as unemployed at the end of 1995. Regional rates of official unemployment, however, reached as high as 23%. Yet, with confiscatory tax levels and an extremely active black market, much income and employment remains under-ground. While it is clear that official statistics do not capture many of the most vibrant aspects of the Russian economy, there is no doubt that life remains difficult for the pensioner, the student and those still dependent on old enterprises. Wage arrears in the state and industrial sectors hit 13.4 trillion roubles ($2.8bn) on 1 January 1996, an increase of 219% over 1994.

These year-end statistics suggest little of the drama that created them or of the drama to come as economic progress is tested by the political manoeuvring between the December 1995 parliamentary elections and the June 1996 presidential contest. The December elections galvanised Presi-dent Yeltsin's survival instincts. The steward of the privatisation pro-gramme, Anatoly Chubais, was forced out of government in January 1996, and was replaced by Vladimir Kadannikov, the former head of the largest Russian automobile manufacturing enterprise. Though largely resisting the emission of credits in late 1995, by February 1996 Yeltsin appeared to see at least the promise of such credits as a key part of his political strategy. He has issued decrees raising pensions and student grants, as well as promising back wages to industrial and state-sector workers. If these promises are followed by actual cash payments, Russia will be in for another round of soaring budget deficits and high inflation.

Despite the threat of these adverse signals, the International Monetary Fund (IMF) agreed in mid-February 1996 on a $10.2bn loan package to help the economy over the next three years. Like the $6.5bn package agreed in March 1995, the loan was linked to Russian promises of lower budget deficits and renewed zeal to control inflation. The IMF dismissed suggestions that the loan was intended to help Yeltsin confront the communist drive for the presidency, but the programme will have delivered about $1.3bn by mid-June. The promised $300 million per month in credits will provide aid that the Yeltsin government can use in the vital months before the June election to help counter the widespread poverty that is one of the more painful consequences of creating a market economy.

Other issues remain to be addressed, such as the stability of the banking system, particularly after the lending crisis at the end of August 1995, the stability of state revenues in light of unsettled centre–region relations, and the lack of basic laws and regulations that would allow outsiders to judge the benefits, costs and risks of doing business in Russia. But the most crucial issue is the uneven distribution of wealth and its political consequences. In fact, in 1995 it became clear that the greatest conflict was no longer between advocates and opponents of reform, but rather between elite groups for state assets and for making the rules that would control their distribution, and between those groups and the rest of society.

Careful examination of the statistics indicates that the benefits of reform and the pain of transformation are unequally distributed. Much more importantly, whatever the actual statistics presented, a broad segment of society believes that the system has produced a few spectacular winners and many losers. The losers, and those who fear they are becoming losers, believe that this unequal distribution is not the result of market forces, but of corruption, collusion between business and government, and unfair practices. They represent a large reservoir of discontent to be tapped in the presidential elections and beyond.

Fighting a War at Home

On 9 December 1994, about 40,000 Russian military and internal security forces invaded Chechnya. While many expected the conflict to evolve into a protracted guerrilla war, few imagined it would last so long. It was nearly two months before a ruined Grozny was captured. Before this victory, Russian forces suffered a serious defeat, as armoured columns became disoriented in Grozny, resulting in serious losses. By July, a shaky cease-fire was in place, but only after extended campaigns against individual villages or mountain strongholds that a succession of Russian spokesmen confidently predicted were the last holdouts of Chechen President Dzhokar Dudayev and his forces.

A nineteenth-century Russian general described Chechnya as 'a fortress' that had to be besieged, and Russia's siege continued into the early months of 1996 with no end in sight. Russian attempts to form a local alternative to the Dudayev regime have failed miserably. Under Russian protection, elections for a new Chechen president were held in December 1995, but the new government has little legitimacy in the republic as a whole.

A military solution appears no more likely than this political solution, and Russian forces show all the signs of profound demoralisation. A mixture of regular army and internal security forces, they favour aerial and artillery bombardment that avoids close encounters with the enemy. Discipline frequently breaks down, with regular reports of atrocities against civilians. The Chechen fighters have grown in confidence as the war has moved into a guerrilla phase, and they choose their targets. They have even taken the war beyond Chechnya itself, seizing a hospital in

Budennovst near Stavropol in June 1995 and another in Kizlyar, Dagestan, in January 1996.

The peaceful resolution of the first crisis, through Prime Minister Chernomyrdin's televised negotiations with the leader of the hostage-takers, was seen by many in the military as a humiliation. But it could not compare to the humiliating and brutal resolution of the second incident, in which Russian helicopters and artillery pounded hostages and hostage-takers alike in Pervomayskoye where the Russian troops had trapped them. The reluctance of the Russian troops to engage directly with the Chechen fighters, however, allowed many of them to escape. In November 1995, Chechens buried canisters of radioactive caesium in a Moscow park and announced this to the world. In January 1996, Chechen sympathisers seized a Turkish ferry on the Black Sea. The successful handling of this incident by Turkish authorities led to the release of the hostages without loss of life. But the return of Chechen fighters to Grozny in mid-March 1996, when they held large sections of the city for four days, further embarrassed the Russians.

In February 1996, Yeltsin expressed fears that his re-election would be in doubt unless the conflict came to an end. With no government consensus on how to resolve the crisis, he placed Prime Minister Chernomyrdin in charge of a solution. This is, at best, an unenviable assignment, more likely to clear the way for Chernomyrdin's replacement in the spring than to resolve this bloody conflict.

Foreign and Security Policy

The fragmentation of the executive branch and the rise of new interests also shaped Russian foreign and security policy. In particular, 1995 was a year of dwindling influence for the Russian Ministry of Foreign Affairs and its head, Andrei Kozyrev. Yet the weak and disorganised policy that ensued was often masked by apparent consensus on key security issues, particularly on the need to maintain and restore Russia's status as a great power. The aim of building on the territory of the former Soviet Union an integrated political, economic and security structure led by Russia was scarcely questioned by anyone.

Russian policy towards the West still retains important elements of cooperation, but the relationship faced a crisis throughout much of 1995 over specific issues such as NATO expansion, nuclear security and proliferation issues, and Bosnia. Both sides maintained an ambitious schedule of bilateral and multilateral contacts and managed to avoid stalemate on key problems. But the long-term trends towards mutual disengagement are clear and unlikely to be papered over by another year of meetings. Moreover, the June 1996 elections could well destroy existing structures of cooperation if the result yields a President Zyuganov or Lebed.

The unresolved question of NATO enlargement dominated Russia's relations with the West. At a December 1994 meeting of heads of state in Budapest, President Yeltsin warned of a 'cold peace'. Russia at first refused to participate in NATO's PFP programme. When it eventually signed the agreement in June 1995, it continued to threaten that the eventual expansion of the Alliance would end cooperation of this type and result in Russia taking various steps in response.

In the autumn, after the September publication of the 'Study on NATO Enlargement', the Russian press featured an extended discussion of possible 'countermeasures'. These focused on everything, from reintroducing Russian combat forces and tactical nuclear weapons in Russian and Belarusian regions adjacent to the NATO area, to Defence Minister General Pavel Grachev's threat of forming a rival alliance with China. Various official and unofficial commentators have threatened to unravel arms-control agreements such as the 1992 Conventional Forces in Europe (CFE) and Strategic Arms Reduction Talks (START) Treaties, or to deepen military cooperation among the CIS states.

Most of these threats have been carefully couched in conditions that reserve them not simply for NATO expansion, but for the introduction of offensive forces or nuclear weapons into the new member-states. But they indicate a strand of thinking that would destabilise relations between Russia and the West. Russian–Belarusian military agreements in December 1995 – and the prospect for broader political integration in 1996 – create the basis for Russian moves that could reverse the nearly decade-long trend away from concentrating significant military forces in Eastern and Central Europe.

On Bosnia, senior Russian officials were staunchly opposed throughout the year to NATO's bombing campaigns and domination of the peace process, climaxing with Yeltsin's statement in September 1995 that NATO bombing of Bosnian Serbs fanned 'the flame of war all over Europe'. In the end, the two sides found a formula in November for Russia's participation in the peacekeeping mission in Bosnia. This included Russian deployment of a small force of 1,500 troops under US, but not NATO, command. This outcome clearly owes much to the negotiating skills of both sides, but equally reflects Russia's inability to send more than a token force, combined with its unwillingness to sit on the sidelines as NATO troops, and troops of would-be NATO members, deployed in Bosnia.

Arms-control and non-proliferation questions provided some of the most difficult problems in 1995. Long-standing Russian objections to flank limits in the CFE Treaty led Russia to warn that it would ignore these limits when the Treaty entered into force in November 1995. NATO responded before this deadline by opening serious talks with Russia to produce a compromise. Though November passed without Russian compliance, a compromise appeared to be in the offing that would preserve the Treaty and yet give Russia some relief on the flanks.

The START II Treaty, concluded in January 1992 and ratified by the US Senate in January 1996, was still unratified by the Russian Duma, and sceptics doubt that the Duma will act on it during 1996. However, financial pressures are already creating a serious imbalance between US and Russian nuclear force levels that could well make the Treaty more attractive to Russians concerned with maintaining nuclear parity with the US.

The safety and security of Russia's nuclear forces and materials remained a strategic focus of Western initiatives, as did efforts to head off Russian construction of light-water nuclear reactors at Bushehr, Iran. In January 1995, Russian Minister for Atomic Energy Victor Mikhailov visited Tehran and agreed on a deal that would net Russia close to $1bn in return for completing construction on nuclear reactors at Bushehr. The issue remained unresolved, despite intensive efforts by senior Western leaders throughout the year to find a compromise. It was perhaps the best example of the fragmentation of the Russian foreign-policy process. There was considerable sympathy in some Russian quarters for curtailing the deal, but financial pressures and the independence and power of the Ministry of Atomic Energy made such an outcome problematic.

Russian–Chinese relations were bolstered in 1995. In June, Chinese President Jiang Zemin visited Moscow, culminating a long series of visits by negotiating teams, arms-export teams, and senior defence and foreign ministry officials. The Chinese and Russians sounded a note of accord in rejecting what they see as US interference in their internal affairs, saying that they would 'not allow anyone to teach us how to live and work'. However, it takes more than such moments of ideological solidarity to provide a firm basis for an alliance. The Chinese continue to buy Russian arms, but unanswered questions regarding China's future stability, its long-term ambitions in Central Asia and cross-border migration in the Far East make for an unsteady partnership at best.

Russia and its Neighbours

Integration with the states of the former Soviet Union remained a key aim of Russian foreign policy. In September 1995, Yeltsin signed a decree defining Russia's vital interest in integrating within the CIS. Both before and after the decree, regular CIS and bilateral meetings addressed an ambitious agenda of economic, political and security ties. Moscow has offered military and economic cooperation to lure the former Soviet republics into integration. In November 1995 and January 1996, CIS leaders agreed to a joint air-defence system by creating air defences in Georgia, Kyrgyzstan and Tajikistan, and by improving those of Armenia, Kazakhstan and Uzbekistan. Russia will bear the economic burden of any improvements and will effectively control the unified system. In addition, Russia, Belarus and Kazakhstan signed a payment and customs union agreement in January 1996, anticipating that three other CIS states – Uzbekistan, Tajikistan and Kyrgyzstan – would shortly sign the pact.

But the Kremlin's approach has hardly been one of all carrot and no stick. Any sort of military union under Moscow's control automatically means that the participating states will lose a certain amount of sovereignty. Furthermore, energy supplies from Russia to the former republics is one stick that the Kremlin could use to cajole the CIS members into a reconstituted union. At the November 1995 CIS meeting, Russian Prime Minister Chernomyrdin highlighted the Commonwealth's 14 trillion rouble debt for Russian energy deliveries. His solution to the problem was a payment and customs union. The energy question is of vital importance to Ukraine, which relies on Russia for most of its fuel supplies.

While Moscow has relied more on the 'charm offensive' than on any threats to achieve its goals, its efforts continue to have only limited success and appeal. Moreover, such a policy comes up against the hard realities of this post-Soviet are. Deep integration may be possible with an inner core of states, including Belarus, Kazakhstan and other floundering republics such as Tajikistan. But success with this inner core deepens the scepticism and independence of states such as Ukraine, which stand at the outer edge of the CIS, but nevertheless remain vital to Russian interests.

Moscow and Kiev reached some agreement in bilateral talks during 1995, although Kiev is still sceptical of Russia's integrationist policies. In March, an IMF-inspired settlement of Ukraine's debt was reached. The two presidents also announced the settlement of the Black Sea Fleet issue in June, although, as before, the key details remained unresolved. At the same time, Ukraine has been careful to reaffirm its independence. In July 1995 it signed a treaty of friendship and cooperation with Belarus; in contrast to Belarus' more integrationist moves with Russia, however, no effort was made to establish a customs union. Furthermore, with the Kremlin's attention focused on Chechnya, Kiev reasserted its control over the separatist-leaning republic of Crimea by suspending the peninsula's constitution and ejecting its pro-Russian president from office. Despite such successes, Ukraine continues to exercise caution in dealing with the Russians. The possibility of a President Zyuganov, and a subsequently more concerted Russian effort to deepen integration (and pay the costs), alarms Ukrainian leaders. But until, and if, either comes about the two countries can be expected to move slowly towards a new and careful relationship.

Of major concern to Moscow are the various armed conflicts in the 'near abroad'. Some headway was made in Tajikistan, when the government and rebel representatives actually sat together round the negotiating table in Moscow in the spring. Heavy fighting had broken out earlier in the year, with Russian troops supporting those of the Tajik government and managing to quiet the situation. But the future effectiveness of the Russian-led CIS peacekeeping force there has to be questioned.

While large-scale clashes are now a rarity, Tajik rebels and Afghan *mujaheddin* fighters regularly cross the border from Afghanistan and carry out hit-and-run attacks on Russian border guards and other targets. While these forces show no signs of tiring, CIS resolve for the peacekeeping operation has began to decline. Kazakhstan and Uzbekistan announced in May that they would not contribute their troops indefinitely to the CIS force. Commonwealth leaders in January 1996 were able to extend the peacekeeping mandate until 30 June, but they also put pressure on the Tajik leader, Imamali Rakhmanov, to reach an agreement with the opposition. With the high cost of the war in Chechnya and the Kremlin's pre-election handouts, Russia, in the words of Boris Yeltsin, 'can't hold Tajikistan by the hand'.

There was perhaps greater success registered in 1995 in talks between Armenia and Azerbaijan over the disputed Nagorno-Karabakh enclave. Although clashes have continued to occur and no major peace accord has been signed, the two warring parties held seven rounds of negotiations under the auspices of the Organisation for Security and Cooperation in Europe (OSCE's) Minsk group. In a more encouraging sign, in late December 1995 Armenia and Azerbaijan met in The Hague for bilateral talks without a mediator. As a result, the May 1994 cease-fire has generally held and there is optimism that a mutually agreeable solution will be found.

To its neighbours, Russia may well present threats that arise from both its strengths and its weaknesses. This potentially dangerous combination could appear as Russian power flows into the weak and unstable areas of the new borderlands, even as it is rebuffed in Europe and East Asia. The future of Eurasia and Russia's role in it is likely to turn on whether Russian leaders of whichever political orientation can adapt their image of the country to its real capabilities, or whether they will try to make those capabilities fit some exalted, but unattainable image. The June 1996 election is an important turning-point, but it will not provide the definitive answer. Whoever wins will have substantial powers to shape Russia's future, but will not exercise a monopoly on power. And, no matter how clear his mandate might appear, his resources are unlikely to be as great as the problems he will face in the years ahead.

A Fragile Peace For Bosnia

After four years of bloody war, a General Framework Agreement for Peace in Bosnia-Herzegovina was signed by Bosnian President Alija Izetbegovic, Croatian President Franjo Tudjman and Serbian President Slobodan Milosevic at the Elysée Palace in Paris on 14 December 1995. As the protagonists in the war of the Yugoslav succession shook hands on the podium, public attention focused on the contribution that diplomacy and high-profile mediators had made towards securing a settlement. While the settlement had required patient diplomacy and decisions reached at the negotiating table, a more significant role had in fact been played by actions on the battlefield.

There was more fighting, suffering and ethnic cleansing in Bosnia and Croatia in 1995 than at any time since the early stages of the war in 1991–92. By early October, changes in the balance of forces on the ground – first in Croatia and later in Bosnia-Herzegovina – had created a radically different, and more conducive, context for the US-led diplomatic initiative launched in mid-August to produce a comprehensive peace settlement. After three weeks of intensive and secretive negotiations, agreement was eventually reached in Dayton, Ohio, on 21 November. The precise nature of that agreement, however, and its implementation since it was signed in Paris, have inevitably raised questions about the long-term prospects for peace in the Balkans.

The Unsafe 'Safe' Areas

On 17 March 1995, General Rasim Delic, Commander of the Bosnian Army, observed that the cease-fire brokered by former US President Jimmy Carter in December 1994 had done little more than enable the Bosnian Serb military to fortify its positions and consolidate its hold on nearly 70% of Bosnia's territory. Earlier that month, Croatian Army commanders, the Bosnian government and Bosnian Croat forces (HVO) had agreed to a formal military alliance. With little evidence of progress round the negotiating table, it did not come as much of a surprise when some 2,000 Bosnian government forces resumed military operations on 19 March 1995 by launching an attack north and east of Tuzla against Serb positions in the Majevica mountains.

At the same time, on Mount Vlasic near Travnik in central Bosnia, government troops launched an attack that quickly put Serb forces on the defensive. Throughout April, well-trained and better-equipped government forces maintained the initiative, capturing strategically important communications facilities on Mount Vlasic. Along the mountainous Croat–Bosnian border, HVO forces, supported by regular units of the Croatian Army, also captured territory, thus putting them within artillery range of Knin, capital of the Serb-held Krajina region of Croatia. Towards the end of April, Bosnian government forces also began to recapture territory south of Bihac, the government-held enclave in north-western Bosnia where fighting had continued more or less uninterrupted since the December 1994 cease-fire.

As in the past, the Serbs responded to Bosnian government military operations by intensifying their indiscriminate bombardment of Bosnian cities, especially Sarajevo and Tuzla. As the shelling increased throughout April and May, it became clear to General Rupert Smith, who had replaced General Sir Michael Rose as the United Nations Force Commander in late January 1995, that the continuation and precise role of UNPROFOR would again soon be questioned.

On 7 May, when a single shell killed 11 people in a Sarajevo suburb, General Smith responded promptly by requesting air-strikes against Serb targets. Yasushi Akashi, the UN Secretary-General's Special Representative for the former Yugoslavia, turned down the request and was sharply criticised for doing so by US Ambassador to the UN Madeleine Albright and the newly appointed French Prime Minister, Alain Juppé. Akashi argued that the bombing would endanger the security of UN peacekeepers dispersed in vulnerable locations throughout Bosnia – such as at weapons collection points and in the UN-designated 'safe areas'. He was also concerned that an attack might affect the already bleak situation in Croatia where hostilities had resumed in the first week of May when the Croats attacked western Slavonia. As a result, UN Secretary-General Boutros Boutros-Ghali, increasingly unhappy about the UN's commitments and ill-defined role in the former Yugoslavia, ordered a 'fundamental review' of UNPROFOR involvement.

It was too late for these second thoughts, however. Internal discussions within the UN and among members of the Contact Group (France, the UK, the US, Russia and Germany) were almost immediately overtaken by events on the ground. On 16 May, a Bosnian government offensive south-east of Sarajevo led to a seven-hour Serb artillery barrage against the city. Over the next week, fighting intensified and, on 24 May, further shelling prompted General Smith to issue a three-part ultimatum to both sides: shelling into the exclusion zone was to cease forthwith; weapons taken from UN collection points were to be returned by noon on 25 May; and all heavy weapons in the exclusion zone were to be withdrawn by 26 May. Failure to comply, Smith affirmed, would lead to NATO air-strikes.

This time there was no bluffing. Shortly after the first deadline expired, NATO aircraft struck an ammunition dump and destroyed two bunkers near the Bosnian Serb stronghold of Pale. For the first time, NATO aircraft had deliberately attacked what the UN military spokesperson described as a 'significant military infrastructure target'. The following day, continued Serb shelling and refusal to remove artillery pieces from the exclusion zone led to a second series of air-strikes, which destroyed a further six weapon bunkers. This provoked a Bosnian Serb artillery attack against Tuzla which killed more than 70 people and left nearly 200 injured.

To counter the threat of further air-strikes, Bosnian Serb leader Radovan Karadzic and General Ratko Mladic ordered their troops to take UNPROFOR personnel hostage and hold them at or near 'potential NATO targets'. By 1 June, the Bosnian Serbs had taken nearly 300 hostages. By parading UN 'peacekeepers' – demoralised and chained to likely military targets – before the world's television, the Pale leadership revealed the UN's impotence.

What might at first have appeared to the Serbian leadership as a clever move, however, quickly backfired. While the hostage-taking was deeply humiliating for the UN and might even have brought a temporary respite from further attacks, it became a turning-point in the war. Along with NATO's decision to abandon its 'pin-prick' approach and to target significant elements of the military infrastructure, the hostage crisis forced a number of decisions which enabled the UN Force Commander, in effect, to shift the operation from peacekeeping to peace enforcement.

Strong pressure from Serbian President Milosevic – using his State Security Minister Jovica Stanisic as a special envoy – convinced the Pale leadership to release all the UN hostages on 18 June. The same day, the UN withdrew all its personnel from the weapons collection points in Serb-held territory. While the concept of a heavy-weapons exclusion zone thus collapsed, UN personnel became less vulnerable to isolated attack and hostage-taking.

As a further consequence of the hostage crisis, UK Prime Minister John Major announced on 28 May that the existing UK contingent would be reinforced by a further 1,200 troops, while promising that another 5,500, including two artillery batteries, would be sent out later. The following week, NATO defence ministers agreed to set up a Rapid Reaction Force under UN command. The proposed force, made up of UK, French and Dutch units, was to include helicopters, anti-tank weapons and artillery batteries. The UN Security Council approved the 'expansion of the peacekeeping force by up to 12,500' on 16 June.

The precise role of this Rapid Reaction Force became the subject of considerable public wrangling between UK and French officials. Even though the debate was never clearly resolved, the Force's very existence made a difference. UK officials in particular might have stressed the 'defensive' purpose of increasing UN troop levels, but the introduction of heavier weapons changed the balance of forces on the ground and broadened the range of responses available to the UN Force Commander.

The Fall of Srebrenica and Zepa

On 14–15 June, in an attempt to break the siege of Sarajevo, the Bosnian Army launched a large-scale attack in the Visoko area. The Serbs, controlling the high ground and with superior fire-power, responded by further intensifying their shelling of Sarajevo. At the same time, military operations against other 'safe areas' in eastern Bosnia were stepped up. On 11 July, with the Contact Group in public disarray over what to do next, Serb forces captured Srebrenica, one of the six areas designated as 'safe areas' by the UN in May 1993. In a belated attempt to end the Serb offensive, NATO launched two air-strikes, although follow-up strikes were halted when Serb soldiers threatened to kill Dutch peacekeepers captured in the initial offensive against the enclave.

With Zepa and Gorazde also under threat, the Foreign and Defence Ministers of the Contact Group and troop-contributing countries, and officials from the EU, UN and NATO set up a conference to convene in London on 21 July. While Western leaders focused their attention on Gorazde as the next 'safe area' on the Serb list, the Bosnian Serbs, Croat Serbs and forces loyal to Fikret Abdic, the pro-Serb Muslim warlord in Bihac, launched a coordinated attack against the Bihac enclave in the north-west. Responding to this new threat, Croatian President Tudjman and Bosnian President Izetbegovic met in Split on 22 June and agreed to intensify military cooperation. As a result, regular Croatian Army forces, reportedly as many as 10,000, entered Bosnia on 27 July and, fighting alongside HVO forces, attacked positions south of Bihac. By capturing the key towns of Bosansko Grahovo and Glamoc, Croat forces cut Serb supply lines; this ominously presaged the much more comprehensive Serb collapse that was shortly to follow.

With the threat to Bihac countered and Western attention still focused on Gorazde, Serb forces overran the enclave of Zepa on 25 July. Evidence soon emerged that Serbian actions were accompanied by serious atrocities. Tadeusz Mazowiecki, the UN representative for human rights in the former Yugoslavia, reported how Serb actions in Srebrenica had involved 'very serious violations of human rights on an enormous scale that can only be described as barbarous'. When similar evidence of atrocities perpetrated against the population of Zepa came to light, Mazowiecki resigned in protest at the lack of a UN reaction.

The Contact Group members had agreed in their final statement to the 21 July London Conference that an attack on Gorazde would meet with a 'substantial and decisive response'. Major disagreements among the five powers, however, were difficult to disguise. As the Conference adjourned, Russian delegates openly rejected large-scale air operations, while US officials argued that 'decisive response' could only mean an extensive air campaign in reaction to further Serb attacks on 'safe areas'. The US position was indeed logical given that the UK had withdrawn its contingent from Gorazde and that a French proposal to reinforce the enclave using an armada of helicopters had been rejected by the US. As UK Foreign Secretary Malcolm Rifkind vainly sought to paper over the cracks, US Senator Bob Dole captured public sentiment by describing the Conference as 'a dazzling display of ducking the problem'.

When Zepa fell, pressure for action against the Serbs intensified. Following a crucial meeting of the North Atlantic Council (NAC) on 25–26 July, Boutros-Ghali agreed to change the 'dual-key' arrangements that had hitherto governed the procedures relating to the authorisation of air-strikes. To ensure a 'decisive response', the final authority was shifted from Yasushi Akashi to the UNPROFOR Force Commander in Zagreb, General Bernard Janvier. Because this transfer of authority technically occurred within the UN mission, no Security Council resolution was deemed necessary, and Russia was therefore unable to veto the change of policy.

Soon after the NAC meeting, the threat of air-strikes was extended to protect the remaining 'safe areas': Tuzla, Bihac and Sarajevo. In another important development, UK and French artillery units of the Rapid Reaction Force moved to Mount Igman between 23 and 24 July. With UN personnel no longer in exposed positions inside Serb-controlled territory, the ground had been prepared for a transition to peace enforcement.

Croatia on the March

Since the end of the Serbian–Croat war of 1991, the Croatian Serbs had retained control of some 30% of Croatia's territory, including the Krajina region and western and eastern Slavonia. The UN had been unable to restore the so-called UN Protected Areas (UNPAs) encompassing these regions to Croat control. This failure, when added to the need to reintegrate some 300,000 refugees displaced by the fighting in 1991, the economic consequences of continued Serb control (especially for the tourist industry on the Adriatic coast) and, above all, fear that the status quo might become permanent, fuelled President Tudjman's growing impatience.

But most importantly, since the signing of the Washington Agreement in 1994, the Croatian Army had been rebuilt and restructured with US aid and expertise. A wholly ineffective arms embargo allowed Croatia to build up its air and mechanised units and to receive substantial training from private US contractors, such as the Virginia-based Military Professional Resources Incorporated (headed by Lieutenant-General Ed Soyster, former Deputy Head of the US Defense Intelligence Agency). By early 1995, Tudjman was ready to try the military option.

President Franjo Tudjman had threatened in early January 1995 to terminate the UN mandate in Croatia once and for all. International concern was immediately expressed about the 'risk of renewed hostilities' and diplomatic efforts to reach a compromise between the self-styled Serb Republic of Krajina (RSK) and the Croatian government were intensified. As fears about a wider war mounted, Tudjman agreed on 12 March to a six-month extension of the UN presence. He may have seemed to have yielded to international pressure, but Tudjman had in fact secured important concessions. The size of the UN force, renamed the UN Confidence Restoration Operation in Croatia (UNCRO), was sharply reduced (from 14,000 to 5,000) and its mandate was modified to include a commitment to international control of the border between Croatia and Bosnia. In any event, Croatia was already preparing for war, and the extension of the mandate was certainly not viewed by Tudjman as an impediment to renewed military action.

The carefully planned offensive began in early May. Some 7,000 Croatian troops moved across the line and quickly captured the vulnerable Sector West in western Slavonia. The significance of *Operation Flash* was not lost on the Croatian leadership. It confirmed that the armed forces had indeed been effectively rebuilt and reorganised under Defence Minister Gojko Šušak. The Croatian Army's improved equipment and effective US training exposed the Serb forces as more vulnerable and far less impressive than they had appeared earlier in the war. Of equal importance, Serbian President Milosevic did nothing to prevent the Croats from retaking Serb-controlled territory. On the contrary, he even appeared to distance himself from the Krajina Serbs by condemning the rocket attacks against Zagreb which Milan Martic, the Croat Serb leader, had ordered in response to the Croatian attack.

The international response to the Croatian attack was, from Tudjman's point of view, highly encouraging. The attack was a flagrant violation of UN-sponsored agreements that had just been renegotiated to meet key Croatian demands. Yet, when the UN Security Council met to discuss the matter in closed session, the US and Germany effectively blocked any resolution or statement that would have formally condemned the attack. The following month, after Croat forces seized the strategic Mount Dinara in Bosnia on 4–8 June, Martic formally cancelled the economic agreement reached in December 1994. Tudjman, emboldened by the success of the western Slavonia operation, responded by threatening further military action.

The Krajina Offensive

On 4 August 1995, the Croatian Army launched *Operation Storm*, a large-scale, carefully coordinated attack on five fronts in the Krajina region. Serbia's defences collapsed rapidly and the Croats recaptured the whole territory in less than a week. One immediate consequence of this was the most serious refugee crisis since the beginning of the Yugoslav conflict. Within days of the offensive, some 150,000 Serbs had fled into Serb-held parts of Bosnia and Serbia itself. Croatian authorities were unable to conceal and unwilling to prevent, the atrocities and widespread human-rights violations carried out during and after the campaign. European Union Monitors (ECMM) and the UN Human Rights Action Team (HRAT) compiled detailed reports documenting Croatian activities, including summary executions, the shelling of refugees, and the systematic torching of Serb homes and villages.

As in the case of western Slavonia, however, the US and Germany were reluctant to condemn Croatia's action. Indeed, Peter Galbraith, the US Ambassador in Zagreb, denied that Croatian activities amounted to 'ethnic cleansing' and US President Bill Clinton, in a statement on 7 August, said he was 'hopeful that [the Croatian offensive] will turn out to be something that will give us an avenue to a quicker diplomatic resolution'. Thus, on 9 August, the US administration launched a new peace initiative for Bosnia based on the existing Contact Group map, first presented to the parties in July 1994. European Union officials also began to see the possibilities that the changing military situation on the ground might offer for a broader settlement. As one diplomat reportedly observed regarding whether the ECMM report on Croatian activities in the Krajina should be made public, 'there's enough of an understanding with Croatia to let sleeping dogs lie'.

In spite of all the criticism of the UN's performance in Croatia between 1992 and 1995 (most notably from the Croatian government), the Krajina offensive did not result in the UN's complete disengagement from Croatia. The organisation still has its uses. On 15 January 1996, the UN Security Council established a new peacekeeping force in the Croatian territories bordering the Federal Republic of Yugoslavia (FRY) still under the control of local Serbs. In accordance with a Basic Agreement reached by the Serb community and the Croatian government on 12 November 1995, a UN Transitional Administration for Eastern Slavonia, Baranja and Western Sirmium (UNTAES) has been established for an initial period of 12 months. Consisting of some 5,000 military and civilian personnel, its task will be to govern the region and oversee its 'peaceful reintegration' into Croatia. With President Milosevic clearly not prepared to intervene militarily to defend eastern Slavonia, the Serb community there is seeking guarantees of equal treatment and maximum autonomy from Croatia, but is ultimately reconciled to incorporation. Any stalling on their part, as the fate of the Krajina Serbs showed only too clearly, may lead to further action by the Croatian Army.

The Beginning of the End

The threat of an extensive NATO air campaign issued after the 21 July London Conference had produced a temporary lull in the shelling of Sarajevo. Moreover, the dramatic collapse of Croat Serb positions in the Krajina forced the Bosnian Serbs to consolidate their positions in Bosnia. They were also forced to cope alone with the massive influx of refugees in northern Bosnia. President Milosevic made clear that he was far more concerned about his own position within Serbia than about the fate of fellow Serbs in either Croatia or Bosnia. General Mladic nevertheless remained defiant, apparently not believing that this time NATO was serious about conducting a sustained air campaign to defend the remaining enclaves.

On 28 August 1995 a shell exploded in the Markale market-place in Sarajevo, the site of the February 1994 massacre which had led to the establishment of the original weapons exclusion zone. It killed 37 people, and UN officials were quick to declare that it had been fired from a Serb mortar position. On 30 August NATO responded with *Operation Deliberate*

Force. Within 12 hours, some 300 sorties were flown against radar, communications, artillery and missile sites throughout Serb-held parts of Bosnia. A pause in the air campaign on 1 September was designed to give Serb military commanders time to comply with NATO demands: an end to the shelling, the opening of routes into Sarajevo and the immediate withdrawal of heavy weapons.

The Serb response did not satisfy General Janvier and he ordered the air campaign to be resumed on 5 September. When the Serbs reacted with further shelling, the Rapid Reaction Force near Sarajevo also opened fire for the first time since its deployment. By 13 September, NATO had carried out 800 bombing missions and was beginning to run out of military targets. The following day, the air-strikes were suspended following a firm commitment by the Bosnian Serbs to pull their heavy weapons around Sarajevo out of range.

The Pale leadership continued to bluster in public, but it was clearly deeply shaken by the air-strikes. Indeed, on the very first day of the NATO bombardment, the Bosnian Serbs announced that they would now join Belgrade in negotiations as a 'joint team' led by President Milosevic. On 8 September, with the air attacks still under way, the Bosnian, Croat and FRY foreign ministers met face to face in Geneva for the first time in 18 months. Under the auspices of Richard Holbrooke, US Assistant Secretary of State for European and Canadian Affairs, agreement was reached on the basic principles for a peace accord.

Critical issues remained, not least how to reach agreement on the territorial divisions while the Bosnian Serbs still controlled some 70% of the country. Once again, developments on the ground pushed the process forward. On 11 September, Bosnian government forces and the Bosnian Croats took advantage of *Operation Deliberate Force* by launching a large-scale offensive in Western and Central Bosnia. Over the next week, Bosnian Serb forces, severely weakened both psychologically and materially by the NATO air offensive, were forced out of 20% of the territory under their control, thus bringing the situation on the ground close to the 51 to 49 division envisaged in the Geneva agreement. On 24 September, 'Further Principles for an Agreement' were reached in New York.

The Dayton Peace Accord

Holbrooke brokered a 60-day cease-fire which began on 12 October 1995. A few weeks later proximity talks began in Dayton, Ohio. The resulting Dayton agreement, initialled on 21 November by Presidents Izetbegovic, Tudjman and Milosevic, was based on the Geneva principles agreed in September. While Bosnia technically retains its legal status as a single state, the parties also recognise the co-existence within that state of two distinct 'entities': the Bosnian–Croat Federation; and the Republika Sprska

– controlling 51% and 49% of Bosnia's territory respectively. Moreover, under the peace agreement, each of the 'entities' is allowed to form 'special parallel relations with neighbouring countries'.

In order to strengthen the integrity of Bosnia as a whole, the peace agreement provides for federal institutions in which the three major ethnic groups are to be represented. These include: a rotating presidency; a bicameral national parliament; a Constitutional Court; and a central bank. The central government, at least in theory, will remain solely responsible for the key areas of foreign affairs, foreign trade and monetary policy. To implement the accord, the UNPROFOR operation has formally been terminated and NATO has assumed the lead in establishing and deploying the multinational peace implementation force, IFOR. This force, numbering nearly 60,000 and equipped 'to meet any contingency', is operating 'under the direction and political control of the North Atlantic Council, through the NATO chain of command'.

In spite of all this, it is difficult not to conclude that the Dayton accord, in the words of one observer, 'may amount to no more than partition, disguised by lawyerly fictions'. Indeed, the accord itself contains provisions that reinforce ethnic politics and are likely to undermine loyalty to federal institutions (such as laws allowing for multiple citizenship and the maintenance of armed forces by each of the 'entities'). Furthermore, the very strict time-frame for implementing key aspects of the accord, most notably the schedule for early legislative elections, makes it doubtful that confidence can be generated among communities, nor does it allow civilian institutions to acquire the legitimacy necessary for reconciliation.

The activities of IFOR since its deployment have done little to dispel concerns about the long-term viability of the settlement. Given IFOR's size and strength, the narrow military provisions of the accord were never likely to pose insurmountable problems. The civilian and political aspects of the Dayton agreement, however, have proved far more difficult to implement. The difficult 'reunification' of Sarajevo in February and March 1996, and the acute tensions that persisted among Muslims and Croats in Mostar, testify to the problems that are likely to remain long after the projected departure of IFOR in November 1996.

The situation in Mostar raises a further issue. The most immediate threat to the Dayton settlement is the health of the Bosnian–Croat Federation. Since its establishment in March 1994 it has failed to gel into a proper alliance, let alone a viable federation. In spite of several attempts in 1995 to strengthen it – including a US-sponsored arbitration agreement signed in Munich in February 1995 and a package of measures agreed on 8 April 1995 to speed the creation of 'genuine' federal institutions – relations between Bosnian Croat and Bosnian government officials and communities remain fragile and tense.

There has also been a near-complete failure to implement the agreement of 10 November 1995 reached by Presidents Izetbegovic and Tudjman during the Dayton proximity talks. This agreement, which includes measures to repatriate refugees and to end the *de facto* division of Mostar, was seen as an important step towards an overall settlement. Considerable responsibility for the lack of progress lies with Franjo Tudjman – who has consistently maintained, in private and in public, that a unified Bosnia is not a viable or 'legitimate' entity – but also with hardline elements in his Cabinet, notably Defence Minister Šušak, and with the Mostar Croats themselves. Partly for this reason Tudjman has been unwilling to apply effective pressure on the Croat community in Bosnia, whose war effort in 1993 he did so much to support. No doubt also for this reason, he chose to promote Major-General Tihomir Blaskic, former Chief of Staff of the HVO, to Inspector General of the Croatian Army on 13 November, the day after Blaskic had been formally indicted for war crimes by the International Criminal Tribunal for the Former Yugoslavia (ICTY) in The Hague. More recently, however, Croatia has indicated it will cooperate with the ICTY and has promised to extradite indicted Croats.

Prospects for Stability in the Balkans

Despite the many problems and challenges that lie ahead, not all developments in the Balkans over the past year have taken a turn for the worse. Nor indeed have all of the UN's efforts been in vain. On 3 October 1995, a bomb blast in Skopje seriously injured the President of the former Yugoslav Republic of Macedonia (FYROM), Kiro Gligorov. This bombing had a salutary effect throughout the southern Balkans. Given its large Albanian minority (23%, according to official figures), FYROM had long been of special concern to regional powers and the international community. Ethnic rioting in early 1995 had underscored the tense relations between FYROM nationals and Albanians, and the threat of full-scale war spreading from Bosnia was a constant worry to governments in Albania, Greece, Bulgaria and the FRY. Securing peace in FYROM was thus seen as the key to peace in the wider Balkans and the first step in preventing yet another war of the Yugoslav succession. To this end, UN forces were dispatched to the Republic in December 1992. Their mandate focused on preventing an invasion of FYROM by the FRY, an unlikely prospect even in the most apocalyptic of scenarios. But the presence of a small, active contingent of foreign troops has had a calming effect on FYROM's turbulent domestic politics. Indeed, the UN Preventative Defence Force (UNPREDEP) mission has succeeded largely because it has never been forced to fulfil its mandate. Instead, UN officials have been free to work on infrastructural projects – building roads, communications facilities, bridges and other structures necessary to buttress FYROM's weak economy – and to carry out low-level shuttle diplomacy between the government and representatives of the restive Albanian minority. This, perhaps, is one indicator that 'mission creep' can be useful.

The Muslim component of Bosnia has not fared so well. Croatia's position in the region, by comparison, has been immensely strengthened as a result of the events of 1995. It has regained control of nearly all of its territory (except eastern Slavonia) and in the process has expelled the Serb population that had inhabited those regions since the sixteenth century. In spite of its human-rights record during and after the Krajina offensive, Croatia also managed to carry out its military operations whilst avoiding diplomatic isolation. The peace settlement for Bosnia was also warmly welcomed by Croatia. It allows for ties between Croats in western Herzegovina and Croatia to be strengthened, and developments in Mostar and HVO-controlled parts of Bosnia since the signing of the agreement do not bode well for the future of the Federation, nor indeed for the viability of the Bosnian state itself.

The Bosnian Serb leadership and President Milosevic also have reason to be pleased with the outcome of the Dayton peace process. For the first time since the conflict began, the Republika Sprska has been accorded a degree of international legitimacy. The territorial concessions obtained by the Serb delegation were more favourable to the Bosnian Serbs than either the October 1992 Vance–Owen Plan or the 'Invincible' plan of 1993 which had envisaged a 'Union of Three Republics'. As for Milosevic's own position of power within the FRY, always a paramount consideration in his calculations, the economic sanctions are gradually being lifted and, as a result, his position in the FRY is being strengthened.

Unless an international effort is made to support Bosnia after IFOR's likely departure in late 1996, it will be the major loser in the wars of the Yugoslav succession. Economic aid for reconstruction is only part of the challenge facing the international community. The Dayton accord has increased rather then reduced the chances for partition. There is thus a need to ensure that Croatia and the FRY do not undermine the Bosnian state from within. It is equally important that Europe and the US keep up at least the appearance of a common policy. Some differences persist, however, for example over whether the Bosnian armed forces should be armed in order to create a new informal balance of power. The peace is very delicately poised as it is. Yet NATO officials and IFOR troop-contributing countries have been reluctant even to discuss the prospect of an 'IFOR-II' after November 1996. Discrete discussions within NATO, however, apparently began in spring 1996. It is already clear, however, that without the presence of a sizeable NATO-led military force to provide enough time for the accord to become effective, it will be difficult to maintain a unified Bosnian state.

The Middle East

Efforts to achieve comprehensive peace in the Middle East had made extraordinary progress in the two years since Israel and the Palestinians reached a tentative agreement in Oslo in 1993. A resurgence of fundamentalist violence, however, now threatens to paralyse the process. The most unexpected act was the assassination of Israeli Prime Minister Yitzhak Rabin by a militant Jewish student, an act that highlighted the serious split that has developed in Israeli society. Israel was also rocked by a series of suicide bombings by Islamic militants that took almost 60 lives.

In reaction, Israeli opinion has shifted away from support for the peace process. At the end of March 1996, opinion hung in the balance. What had looked like an easy victory for the Labour Party in the elections scheduled for May 1996 was thrown into doubt, as was the forward momentum for the peace effort. Should the *Likud* coalition win at the forthcoming elections, it will at best refuse to negotiate any further agreements with the Palestinians, and at worst might try to nudge back the advances already made.

Yet there have been many changes that even *Likud* would be unable to undo. The Middle East has become a very different place from two years ago. At the poignant funeral for Prime Minister Rabin, *keffiyas* mixed with *yamulkas* and the two with US baseball caps as leaders from Arab countries joined Israelis and Western heads of state to say farewell to an acknowledged hero who had died for peace. After the early 1996 suicide bombings in Israel, 12 Arab leaders joined Israelis and Westerners to condemn the killings and promised their help in preventing more. Israel is now recognised as a state by many Arab countries, with others close to doing so, and Arab and Israeli businessmen are beginning to talk of economic collaboration. Although the new Middle East might be frozen in its present state, it will be difficult to change it back to the confrontational region it was pre-Oslo.

Syria has still not committed itself to full participation in the peace process. It has held serious negotiations with the Israeli government, including reviews of the security situation by high-level military officers, but its position is ambiguous. A number of hardline states continue to oppose the process, and the rejectionist states in the Gulf – Iraq and Iran – are completely aloof from it. Neither is in any position to thwart forward movement, however. In Iraq, Saddam Hussein is still in power, but there are signs that his grip is not as strong as it once was. In Iran, there is an ongoing struggle between the militant fundamentalists and the more moderate and pragmatic wing of the ruling party. Iran is suspected of continuing to supply arms and money to the Islamic fundamentalists who are prepared to hurl themselves to eternity if they can kill Israelis at the same time; until these uncontrollable passions can be checked the goal of providing a more secure life for all in the Middle East will continue to recede.

The Bombers And The Peace

For most of the past 50 years, the idea that an Arab–Israeli peace was a real possibility would have been treated with derision. Yet changes in the international environment have led to changes in the Middle East which are slowly, but steadily, transforming all the relations in the area. Three watersheds in the last three years have encouraged those who wish to see the dream become a reality:

- A Declaration of Principles between Israel and the Palestine Liberation Organisation (PLO) was signed on 13 September 1993 in Washington.

- A Jordanian–Israeli Peace Treaty was signed on 26 October 1994, also in Washington.

- An interest in exploring avenues for economic cooperation and integration developed and found expression at conferences in Rabat, Morocco, in October 1994 and Amman, Jordan, in October 1995.

Not everyone in the region is in favour of peace, however. The considerable progress that has been made, and the prospect of even more progress to come, challenges many ideological positions and beliefs that have been deeply held in the Middle East for a very long time. The recalcitrant on both sides have resorted to violence, both to express their rage and in an attempt to halt the peace process. In November 1995, Israel's Prime Minister, Yitzhak Rabin, was assassinated by a militant Jew fanatically opposed to the peace accord between Israel and the Palestinians. In late February and early March 1996, a group of suicide bombers from a militant wing of the *Hamas* organisation brought their battle against the settlement with Israel to the buses and streets of Israeli cities. Aside from the bombers themselves, nearly 60 were killed.

The serious political and psychological repercussions of these events have blocked the peace process at least temporarily, and its future can no longer be taken for granted. The search for a stable peace in the Middle East is shaped and reshaped by the constantly shifting power balance between the regional states and local actors. The leaders who are trying to move from war to peace in the region are challenged by volatile and divided domestic communities. To meet these challenges, they will need unusual political dexterity and extraordinary bravery.

Death of a Hero

Israeli Prime Minister Yitzhak Rabin was felled by an assassin's bullet on 4 November 1995. The shock of his murder was magnified when it became clear that his killer was a Jewish university student. No one could now deny that Israeli society had become deeply divided. Nor could Israelis avoid noting that the main opposition party, *Likud*, and some of its fundamentalist allies, had contributed greatly to the climate of extremism that resulted in Rabin's death.

The loss of Rabin was a serious blow to the prospects for regional peace. Not only was he Israel's Prime Minister and Defence Minister, but he had been a core member of its security and foreign-policy elite for at least three decades. Most Israelis, even those with different views, respected his military record and his character. Thus, the architect of Israel's military victory in the 1967 Six-Day War was able to make political concessions to attain peace without losing credibility. Indeed, the credibility of Rabin's own tough image allowed the Israeli government to win enough domestic support to reach a political deal with the PLO, and also to retain that support despite acts of violence by *Hamas* and the Islamic *Jihad*.

In his last speech, Rabin summed up his position: 'I was a military man for 27 years. I waged war as long as there was no chance for peace. I believe there is now a chance for peace'. In his view, Israel should not, and could not have ignored the opportunities for peace brought about by the fundamental regional and global changes since the early 1990s. In order to seize these opportunities and achieve the twin goal of peace and security, Israel had to take calculated risks, and so it did under Rabin's leadership.

With Islamist violence intensifying throughout most of 1995, domestic support for this high-risk strategy, however, weakened following every new terrorist attack in Israel. Thus the assassination raised the question of whether the Middle East peace process was sustainable in the post-Rabin era. Could it be advanced, or would it be derailed by acrimonious divisions in Israel?

In the immediate aftermath of the assassination the political fortunes of Israel's ruling Labour Party were clearly enhanced. At his funeral, attended by Arab leaders, US President Bill Clinton and other world heads of state, Rabin was portrayed as having lost his own life for the sake of peace. This was particularly apt since his murder came just minutes after he had addressed one of the largest Israeli peace rallies in many years in Tel Aviv. The Labour Party benefited also from the wide-

spread perception in Israel that *Likud* had contributed to the 'verbal violence' that had preceded Rabin's assassination.

Labour also gained from international recognition of Rabin's role and
support for his peace policy, a policy which the new Labour leader and
Prime Minister, Shimon Peres, was committed to follow. Opinion polls
and membership drives by political parties gave Peres and his party a
clear advantage over the opposition. The electoral pact between *Likud*, led
by Benjamin Netanyahu, and the hardline *Tzomet*, led by Raphael Eytan,
did not improve *Likud*'s image or enhance its appeal to the Israeli public.
Divisions within *Likud* caused by rivalries at the leadership level and the
shrinking support for Netanyahu among Moroccan Jews gave Labour an
advantage. This political climate, as well as improved economic conditions – as shown by rising investment and declining unemployment –
were one reason why Peres moved the national elections forward from
October 1996 to 29 May 1996. Peres also recognised that there would be no
major advance in the Syrian–Israeli peace talks in the near future with
which to woo the voters.

Almost as soon as Peres announced the new election date the picture
changed. The campaign of terror launched by *Hamas* in revenge for the
Israel's killing of the most wanted *Hamas* member, Yehya 'the Engineer'
Ayyash, shook Israel. Given the centrality of the security issue in Israeli
society, this wave of violence against civilians shrunk the political advantage that Shimon Peres and the Labour Party had enjoyed early in 1996.
To cope with this threat, the 'vision-gripped' Peres had to demonstrate
that he also had an 'iron fist'. He adopted a cluster of tough positions to
demonstrate that his government had the resolve and resources necessary to deal with the problem.

Stringent security measures were immediately introduced, and Israel's borders were closed to Palestinian workers from the West Bank and
Gaza Strip. Instead of normalisation and open borders between the two
communities there was physical and economic separation. Curfews were
imposed on West Bank villages and towns with a population of 1.2 million
Palestinians. Members of Islamist groups were arrested and their educa-
tional institutions closed, and the houses of the suicide bombers and their
immediate families were demolished. Moreover, Israel threatened to take
further undisclosed action in the self-rule area should PLO Chairman
Yasser Arafat fail to act against the militant Islamists.

Palestinian Elections

Prior to the vicious terrorist attacks of early 1996, Israelis and Palestinians
had managed to maintain incremental progress in settling their protracted
conflict. In September 1995 in Washington, after difficult negotiations, the
two sides signed the Taba Agreement which, one and a half years behind

schedule, set the terms for the second stage of the peace process. The delay was accounted for by the need to accommodate the domestic political pressures faced by the leaders of both sides.

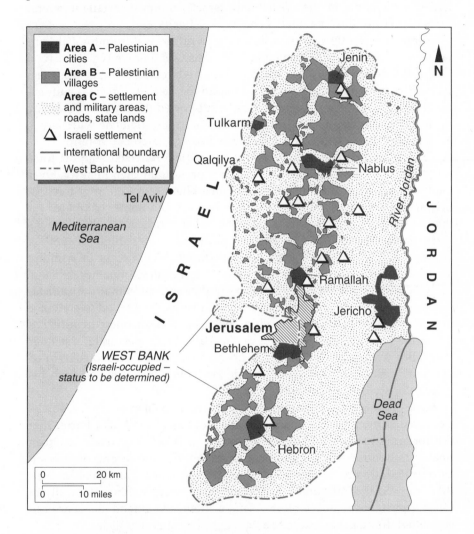

The Taba Agreement covered four areas. In area A, six West Bank towns (Jenin, Tulkarem, Nablus, Qalqilya, Ramallah and Bethlehem) would be placed immediately under the Palestinian Authority (PA's) civilian and military control. Hebron, which has a small Israeli settlement near the town centre, would be under partial PA control, with Israeli forces protecting the settlers. In area B, which includes the majority of the 460 Palestinian villages in the West Bank and contains 68% of the Palestinian population, civilian control would immediately transfer to the PA, while

Israel would maintain control of overall security. Area C, which is mostly made up of rural, sparsely populated hinterland, would be evacuated by Israel as of mid-1996. Israel would retain full control of Area D (the 144 Israeli settlements in the West Bank, Israel's military installations and Jerusalem) until the Final Status end in May 1999.

Israel met the timetable for its military withdrawal from the Palestinian cities in area A on 27 December 1995. This was followed by the arrival of the PA police forces. In addition, hundreds of Palestinian prisoners were released from Israeli prisons. Despite the lack of universal enthusiasm on the part of the Palestinians, most supported the process. They saw no feasible alternative to a negotiated political settlement with Israel, even if they thought that Arafat could have negotiated better terms.

The Palestinian elections on 20 January 1996 reflected support for a settlement and the Palestinians' wish to express their own identity. Over one million of the 1.2 million Palestinians eligible to vote registered in preparation for the elections, and more than 700 candidates competed for the Legislative Council seats. According to most observers, the elections were marred by few irregularities and participation was high (over 80%).

Arafat's victory had never been in doubt. His success in winning about 88% of the popular vote in this first Palestinian election nevertheless enhanced his legitimacy. It was also another indicator that the radical-secular or Islamist opposition to the peace process had failed to convince most Palestinians to boycott the elections. Even in Gaza, often identified as the stronghold of Islamists, 90% of the Palestinian electorate cast their votes. But Arafat did not have it all his own way in the elections; 35 of 88 Legislative Council seats went to independents who disagreed with his policies or style of governance.

One of the written commitments Arafat had given in return for Israel's troop redeployment was to annul those sections of the PLO Charter that call for armed struggle to destroy the state of Israel. Yet Arafat has been unable to honour this commitment since there is still strong opposition among Palestinians to amending the Charter. Those Palestinian groups that opposed the Palestinian Authority, but gained nothing politically in the elections, now want to block Arafat's attempt to amend the Charter to show that they can still exercise a veto over Palestinian policies.

Amending the Charter requires a two-thirds majority in the Palestinian National Council (PNC), some of whose 640 members rank high on the list of those wanted by the Israeli authorities. Arafat attempted to persuade the Palestinian groups headquartered in Damascus to relocate their offices in the territories, but with only mixed success. If convening the PNC becomes difficult, responsibility for amending the Charter may be shifted to the Legislative Council. Some members of the Council favour a conditional annulment by linking the abrogation of the Charter to the release of Palestinian prisoners, the total withdrawal of Israeli troops from Gaza Strip and the West Bank, and the establishment of an independent Palestinian state with Jerusalem as its capital. In effect, this option amounts to changing the terms of the current Israeli–Palestinian peace accords.

If Arafat fails in his attempt to amend the Charter, Israel is likely to retaliate, for example by delaying its withdrawal from Hebron and hardening its negotiating position in the next round of talks. The Charter issue gained political significance in Israel in light of the forthcoming national elections. *Likud* will probably use Arafat's failure to amend the Charter to attack the credibility of the Labour Party's overall strategy regarding the peace process. Concern that the Israeli–Palestinian track would become completely frozen prompted some Palestinian leaders close to Arafat to suggest formulating a new charter to supersede and, in effect, annul the current one, thus avoiding the tensions that are bound to surround its amendment. This alternative is not satisfactory to Israel.

The Final Status talks, which were due to start in May 1996, will probably be slowed down by the elections. Unless some advance is made on the Charter question, a stalemate bewteen Israel and the PA will develop and the talks could be postponed further. If they do take place, the talks will determine the nature and the terms of the final settlement between the two sides. They are expected to last some three years and will deal with the most difficult and contentious issues: the future of Jerusalem; the settlements in the territories; security arrangements and borders; and the future of Palestinian refugees. Given the complexity and sensitivity of these issues, the two parties decided from the outset to postpone talks until more momentum and trust had been achieved. Aside from the intrinsic difficulties of resolving such issues, negotiations will be affected by the national elections and the formation of a new Israeli government. If *Likud* were to win, or if Labour wins but the attacks on civilians continue, there may be no Final Status talks at all.

Arafat and *Hamas*

Israel has had problems controlling the fundamentalist organisation, *Hamas*, and so too has Arafat. He tried to persuade it to participate in the Palestinian elections in return for ending its military operations against Israel and accepting the commitments Arafat made as part of the 1993 Oslo agreement. The *Fatah–Hamas* talks in Cairo in mid-December 1995, however, failed to produce an accord on these issues, although the two sides did agree to avoid military clashes that could escalate into a fully-fledged Palestinian civil war. Some *Hamas* members in the West Bank and Gaza Strip were inclined to form a political party that would compete in the elections. Other *Hamas* leaders, mostly from outside the territories, and

Hamas' military arm, the *Izzedin Al-Qassam* battalions, vehemently opposed participating on the grounds that this would legitimise Arafat's leadership and his peace agreement with Israel. Intense pressure from them convinced the more moderate wing to withdraw its candidacy for the Palestinian Council less than three weeks before the elections.

The radicals favoured escalating the devastating suicide operations in Israel to undermine both Peres' and Arafat's political fortunes. They perhaps hoped that Israel would retaliate in a massive and indiscriminate way, creating greater Palestinian resentment against the Israelis and the PA which cooperates with them. This wing within *Hamas* does not expect the PA to solve the Palestinians' pressing economic problems. Its attacks are intended to make conditions worse by provoking the Israelis to close the borders in retaliation. The level of unemployment in the West Bank and Gaza remains high (around 40%) and frequent closures of the borders between the Palestinian territories and Israel create even bleaker economic conditions. Although much foreign aid has been promised, the amount actually delivered has not been adequate even to cover the PA's budget deficit. *Hamas* also anticipates that the Final Status talks will collapse, thus justifying its opposition to Arafat's deal with Israel. This uncompromising stand will also enhance its popularity with the frustrated Palestinian masses.

Others in the Palestinian Islamist opposition have warned against expecting Arafat's regime to fall because of economic difficulties. They argue that Arafat and the Israelis have created political facts and practical arrangements on the ground which cannot be ignored or overcome through 'armed struggle' alone. After the Palestinian elections, some *Hamas* cadres in the territories signalled their interest in participating in the forthcoming local elections, even though these elections will be run by the PA whose legitimacy has been questioned by the *Hamas* leadership.

The pragmatic wing of *Hamas* already participates in the elections of professional and student associations. It is now trying to broaden its support base by participating in the political process. Believing that military efforts to undermine the settlement with Israel have failed, so far, to achieve their declared objective, it has continued to talk to the PA and, before the last wave of violence, was able to use this political channel to secure the release of more *Hamas* prisoners. It has also been able to convince the Palestinian Authority to allow it to publish an Islamist newspaper and to open an office in Gaza.

Israeli–Syrian Talks

While advances in the Palestinian–Israeli talks have been hesitant, the Syrian–Israeli talks have been painstakingly slow. To achieve a breakthrough the parties must make decisions on issues such as withdrawal, security, settlements, water, and normalising relations. None of this would be easy at any time, but trying to reach such agreements when neither side has any confidence in the other's intentions has proved almost impossible. Despite intense efforts by the United States, the protagonists appeared little closer to a peace accord at the end of 1995 than they had at the beginning.

Washington devoted considerable effort in 1995 to the search for a settlement, believing that such an accord was necessary to reach wider peace agreements in the Middle East, and to normalise relations between Israel and the Arab countries, with the exception of Iraq, Sudan and Libya.

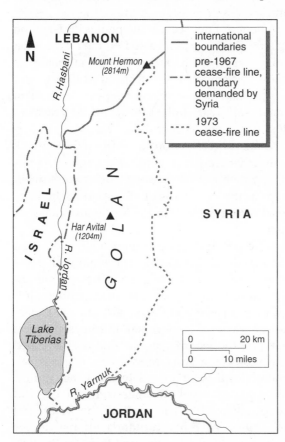

A political deal between Israel and Syria could also consolidate the Palestinian–Israeli agreement by ending Syria's support for Palestinian rejectionist organisations. The Assad government allows rejectionist groups, Islamist and leftist, to have their headquarters in Damascus. Because any agreement would require Syria to collaborate in weakening *Hizbollah*, a peace accord between Syria and Israel would also weaken the Syrian–Iranian alliance.

The US was encouraged by changes introduced by Syria in 1995 to some of its previously adamant positions. It accepted direct bilateral negotiations with Israeli officials without a prior commitment from Israel to withdraw from the Golan Heights. Some of these negotiations, such as the one in June 1995, included discussions between the Chiefs of Staff of the two countries, General Hikmat Al-Shihabi and General Ehud Barak. Syria adopted a new rhetoric for dealing with Israel, suggesting that it was approaching the problem of a settlement in a different way. Under the slogan 'achieving the peace of the brave',

Syria committed itself to pursue a comprehensive peace with Israel which it described as Syria's 'strategic choice'. The Syrian media dropped their usual vitriolic campaigns against the 'Zionist entity' and have taken to referring to the country simply as 'Israel'. They also appear to be identifying major ideological and policy differences in Israel which were either ignored before or treated as signs of a division of roles among Zionists sharing similar values and beliefs. Rabin's assassination provided Syrian leaders with a dramatic illustration that their image of Israel needed fundamental reassessment.

Despite these changes in Syria's positions, optimism about a settlement between the two countries was premature. Although it had lost its Soviet ally, the Syrian government did not see a settlement with Israel on the latter's terms as unavoidable. Syria's leaders acted on the assumption that they still held the ultimate card with regard to a regional peace, as well as to any Israeli–Lebanese settlement. Their decision in 1991 to join the United States and the oil monarchies opposing Iraq provided them with useful cards in the post-Gulf War diplomatic game to reach an Arab–Israeli settlement. Most important among these was stronger Syrian control over Lebanon where *Hizbollah* forces continued to attack Israeli troops and their South Lebanese Army allies. Although Syria has faced considerable economic problems, its domestic troubles were much less acute than those of Egypt in the mid-1970s or of Jordan in the mid-1990s, which had prompted Cairo and Amman to accelerate their exit from the Arab–Israeli conflict.

In terms of Israeli domestic politics, an agreement with Syria that entails an Israeli withdrawal from the Golan Heights is fraught with problems. Israeli leaders have long considered the Heights of strategic significance, leading the Begin government to annex them in December 1981. *Likud* continues vehemently to oppose any withdrawal from the Golan. Labour may be more willing to accept a withdrawal, but insists that its extent can only be determined by the security regime that is established and a full normalisation of relations with Syria. Moreover, a major Israeli withdrawal would be particularly difficult to sell to the Israeli public unless Syria was willing to accept high-level meetings with Israel, as did Egypt and Jordan. Syria simply wanted to win back all the Golan and improve its relations with the United States in return for a minimal degree of interaction with Israel. Accordingly, Rabin favoured freezing the Syrian–Israeli track until Damascus specified the type of peace for which it was ready. Even if a peace agreement with Syria was reached, the probability that it would require considerable withdrawal from the Golan led Peres to promise a referendum to ensure there was public support for such a deal. In the event this was not necessary, for the talks between Israel and Syria stalemated in early 1996.

Normalising Relations?

Bringing wars to an end, even on the regional level, will not lead to genuine peace in the Middle East. Full normalisation of relations is a key component for peace. Such normalisation has to go beyond a 'cold peace' to include such positive elements as mutual recognition, diplomatic relations, economic exchanges, joint projects, open borders, a culture of peace, and communication links between societies as well as states. Shimon Peres' well-publicised vision of a new Middle East in which the pursuit of economic development and regional integration replaces the cult of military strength and power politics represents one such position.

Yet there is continued resistance to normalisation on both sides. While many in Israel support normalising relations with Arab states, they favour complete separation from the Palestinians in the West Bank and Gaza for security reasons. Others consider that Israel is culturally, politically and economically a European country located in the Middle East. On the Arab side, many fear Israeli economic dominance. They argue that Israel's economy is more developed than the Arab economies, which will be turned into markets for the distribution of Israeli goods and sources of needed capital and raw materials. Under these conditions, they warn, Israel will become a regional superpower able to dominate the area both militarily and economically. This in turn could translate into political domination.

The best example of a cold peace is that between Israel and Egypt. There is no fighting between the two states, but Egypt keeps relations between them to a minimum. Egypt's intelligentsia remains either sceptical or opposed to further normalisation. The professional associations, mostly dominated by Islamists, banned their members from visiting Israel and from cooperating with their Israeli counterparts, and the government has not encouraged its citizens to change their attitudes towards the Jewish state. Some Egyptian policy-makers expressed concern that a swift normalisation of relations between Israel and other Arab states might actually develop at the expense of Egypt's regional role. In its confrontation with the Islamists, the Egyptian regime sought to enlist the backing of the Nasserists in the media despite their strong anti-Israeli positions.

In contrast, Jordan and Israel are committed formally to a 'warm peace'. But among the Jordanian people there is still considerable suspicion and mistrust. As in Egypt, the middle-class syndicates dominated by Islamists and pan-Arabists harbour those who argue that real normalisation will lead to Israeli economic hegemony since Israel's economy is more than 12 times the size of Jordan's. Faced with mounting agitation, the Jordanian regime has been clamping down on its critics and has threatened more restrictions in the future. King Hussein's hope is that tangible economic development will gain popular support for his policies, lessen the need for repression and become the cement that secures a lasting peace.

Normalisation on the Syrian–Israeli track is more problematic since levels of mutual suspicion and distrust, if not outright hostility, remain high. Many in Syria would like to end their country's protracted conflict with Israel, but are not ready to go beyond merely ending the state of war. The Arab nationalist ideology of the Ba'ath Party limits the pursuit of normalisation and makes it necessary to move incrementally. From Israel's perspective, Syria continues to shelter Palestinian rejectionist organisations and fails to prevent *Hizbollah* from launching attacks against Israel. Syria's leaders continue to boycott the multilateral peace talks and refuse to engage in the dramatic diplomacy needed to bolster the Israeli public's confidence in Syria's desire for coexistence. Many Israeli analysts believe that Syrian leaders prefer to engage in verbal moderation directed at the US and the West with whom they want closer economic relations in the post-Soviet era.

Nor is there any longer the financial incentive for a peace dividend. Compared to Egypt and Israel, which have between them received $5.5 billion annually in US aid since their peace, Washington now only promises newcomers to the peace process that it will try to convince Japan and Europe to share the bill. The political climate in the United States is much less receptive to assuming responsibility for financing Syrian–Israeli or Palestinian–Israeli agreements. Even waiving Jordan's debts to the US following its peace treaty with Israel was difficult to pass through a reluctant US Congress. Arab oil producers are financially stretched and many have joined the ranks of the borrowers since the end of the 1991 Gulf War. As a result, before the Amman economic conference met in October 1995, they resisted US pressure to provide part of the $5bn capital for a Middle East Bank for Regional Development to be based in Cairo.

Despite the many difficulties, however, there have been some important improvements in relations between Israel and Arab states. Israeli leaders met with 12 Arab leaders at the Summit of the Peacemakers in Sharm al-Sheikh, Egypt, on 13 March 1996. Israeli Prime Minister Peres was invited to visit Qatar in mid-April 1996, and the two countries are likely to develop some form of diplomatic relations soon. Israel has exchanged diplomatic representatives with Morocco and Tunisia, and has made trade agreements with Qatar and Oman. Israeli and Moroccan officials and businessmen examined available avenues of increased cooperation and investment in agriculture. Egyptian President Hosni Mubarak made his very first visit to Jerusalem to attend Rabin's funeral where he was joined by King Hussein of Jordan and the Prime Minister of Morocco, as well as the Foreign Ministers of Oman and Qatar. Reportedly, several Arab countries without formal ties with Israel had also conveyed their condolences to the Israeli Embassy in the US. The Arab boycott against companies doing business with Israel is increasingly less observed, and even the primary boycott is crumbling. Bilateral agreements have been signed between Jordan and Israel in the areas of tourism, agriculture, trade, fighting crime and environmental protection, and there is increasing cooperation between them on telecommunications and transport, with the leaders of these two countries holding periodic meetings.

One feature of Arab–Israeli normalisation efforts during 1995 was the increasing involvement of the private sector in exploring the prospects for economic cooperation. The October Amman economic conference brought together officials and business leaders from most Middle Eastern and North African countries, as well as from Europe, the US and Asia. While policy-makers in countries like Egypt may feel that normalising relations with Israel must wait for a comprehensive settlement, representatives of their business communities have taken part in regional meetings to seek opportunities to establish joint projects. In late February 1996, a large delegation representing the Association of Egyptian Businessmen met with Israelis to discuss possible future economic cooperation between the two countries. They also discussed establishing a 500-member trade council to encourage trade and facilitate joint projects in agriculture and technology transfer.

Although difficulties continue to plague Arab–Israeli negotiations over water, on 13 February 1996 in Oslo, Israel, Jordan and the Palestinians initialled a Declaration on Principles for Cooperation on Water-Related Matters as part of the programme adopted by the Multilateral Working Group on Water Resources. Although complex issues remain to be solved in the context of regional negotiations, this Declaration has defined the main principles guiding their trilateral efforts for cooperation on developing additional water resources, and in identifying suggested areas of cooperation between the parties. The three parties involved also called on both Syria and Lebanon to adhere to the broad principles of the Declaration.

Despite the need to complete negotiations to engage Syria in a more comprehensive peace and resolve outstanding Palestinian issues, there is little question that a new Middle East has come into being. It is far from the vision that Peres has enunciated, but it is light years from the rigid, backward-looking outlook of the remaining rejectionists on both sides. Should Israel's Labour Party win the election in May 1996 the peace process is likely to resume its forward march, however unsteadily. Even if Labour should lose, however, the progress that has been made to date and the changes that have taken place will be difficult to reverse. Should *Likud* come to power it can be expected to halt any further progress, but it would have neither the will nor the ability to change what has already been accomplished.

Egypt: The Struggle With Militancy

In its struggle with the radical Islamists determined to overthrow it, the Egyptian government has been faced with a difficult choice between confrontation and compromise. This dilemma reached a turning-point in 1995. In June, militant Islamists attempted to assassinate President Hosni Mubarak who was in Addis Ababa for the summit meeting of the Organisation of African Unity (OAU). Tensions mounted swiftly following this failed attempt as the Egyptian government moved forcibly to reduce the perceived threat from radicals to its stability.

The attack against Mubarak raised questions inside and outside Egypt about the leadership succession and its impact on domestic stability – particularly as the vice-presidential post had been left vacant since October 1981 when President Anwar Sadat was assassinated and Vice-President Mubarak succeeded him. With no designated successor currently in place, the military may well feel it necessary to seize political power if the presidency suddenly became vacant. The adverse implications of such a development for the prospects of liberalising the Egyptian polity would be far-reaching. These issues were widely debated during the campaign leading up to Egypt's parliamentary elections in late November 1995. They directly influence the hope expressed by many that allowing more political participation might alleviate tensions in society and build more consensus on how to respond to the challenges associated with economic reform.

The Egyptian regime can choose from two models in its battle with radical Islamists, although their clarity was blurred by developments in 1995. On the one hand, there is the model provided by the military regime in Algeria. It had opted for violent confrontation to eradicate the Islamist threat, but later offered some political compromises. On the other hand, there is the approach taken by the Jordanian regime. It had first attempted to co-opt its Islamists into the political establishment by offering them places in parliament, but it later concluded that it had to impose more state restrictions. A combination of political compromise and selective repression is the policy most Middle Eastern regimes, including Egypt, are likely to follow in seeking to deal with the political challenges posed by diverse Islamist movements.

The Islamist Challenge in Egypt

With the exception of the assassination attempt on Mubarak, the level of Islamist violence in Egypt declined considerably during 1995. This hardly meant that the problem was solved, however. The two main militant groups, *al-Jama'ah al-Islamiyah* (the Islamic Group) and *al-Jihad* (Holy War), launched a series of operations inside and outside Egypt to remind the government that its promise to put an end to Islamist violence would not be easy to fulfil.

Most of the violence took place outside Cairo and Alexandria, the two cities targeted by the security services as a top priority. The militants in Upper Egypt, particularly in Al-Minya and Asuit, continued to direct their fire against three targets. The police forces and the civilians who cooperate with them have been at the top of the list, and in January 1996 President Mubarak honoured the families of about 100 victims of such attacks. The militants see these attacks as necessary to deprive the state of what they deem to be its main foundation, its *hybah* (the sense of awe it engenders among the people). Tourist sites and banks were targeted to undermine the economy and to shake whatever confidence the international community has in it. Members of the Coptic community and Coptic-owned businesses were attacked primarily to seize money needed to finance the militant groups and to demonstrate the regime's inability to extend its authority into the reaches of upper Egypt.

In response, the regime used four policy instruments to step up its confrontation with militant Islamists and prevent them from destabilising the status quo. First, the security machinery, particularly the Central Security Forces, run by the Ministry of Interior became involved. The regime hinted that it might even use specialised military units if threats to Egyptian national security continued to emanate from Sudan. Second, the Ministry of Information's controlled mass media were used to portray Islamist militants as obsessed with violence, in sharp contrast to the society's peaceful predisposition. The regime also tried to portray the Islamists as responsible, by disrupting tourism, for the country's socio-economic hardships. Third, the Ministry of Religious Endowments and organs of the religious establishment attempted to show that Islam and violence were irreconcilable and to prove in religious consciousness-raising campaigns that the leaders of militant groups had neither a correct understanding nor much knowledge of Islamic teachings. Fourth, the regime moved to close off the threat from abroad. In particular, it launched a campaign to isolate the Sudanese government on the African and international levels after accusing it of masterminding acts of violence against Egypt. This campaign escalated after Sudan refused to hand over three suspects who had fled to Sudanese territory after the failed assassination of Mubarak. The intensely strained relations between Egypt and Sudan led to an eruption of military clashes along their borders in the contested Halaib area.

When Sudan then seized Egyptian-owned buildings in its capital, Mubarak ordered the immediate eviction of Sudanese soldiers from the border area. Sudan, in response, threatened to reconsider its treaties with Egypt on sharing the Nile river waters. Although it is not clear that Khartoum would be able to control the flow of the Nile water to Egypt, the symbolic significance of the threat triggered harsh responses from Cairo. After Egypt formally accused Sudan of running 50 camps for training militants, including the 'returnees from Afghanistan', and of supporting *al-Jama'ah al-Islamiyah*, Cairo lifted its existing exemption of the Sudanese, requiring all those who henceforth wished to enter Egypt to obtain a visa. In coordination with Ethiopia, Egypt then moved the pressure to the UN Security Council which condemned the Sudanese government and threatened further punitive action if Khartoum did not change its position on the suspects.

The Egyptian government also continued its efforts to reach security agreements with other states to hand over Islamists accused of involvement in acts of violence. In addition to Pakistan, Yemen agreed to return the Egyptian militants who had come from Afghanistan and to expel others so they could no longer use Yemeni territory as a training ground and launching pad for operations against Egypt. The Egyptian regime also tried, though with no success, to get the UK government to expel Egyptian members of Islamist militant groups who had been granted political asylum in the UK.

The government's strong-arm tactics have succeeded in keeping the radical Islamists at bay outside the main cities. Despite intermittent attacks by militant Islamists in upper Egypt, against the Egyptian diplomatic mission in Islamabad, and on an Egyptian diplomat in Geneva, which have all caught press attention, they are clearly not able to overthrow the regime. Like their militant counterparts in Algeria, their campaign of violence failed to gain either popular support against the ruling regimes or even to soak up moderate Islamists.

On the contrary, the campaign has actually backfired. The regime used it to weaken whatever support the radicals might have had and to create more divisions within the Islamist ranks in general. The regime exploited the acts of violence by strongly suggesting that a failure to provide Egypt with much-needed economic assistance could undermine regional stability and bring to power anti-Western political forces opposed to settling the Arab conflict with Israel.

Tarring the Muslim Brotherhood

The confrontation between the Egyptian regime and militant Islamists extended to the regime's relationship with the Muslim Brothers (*Ikhwan*). The Muslim Brotherhood is the most important Islamist opposition group in Egypt. It was established by Hassan Al-Banna in Ismailia in 1928 and became a major actor on the Egyptian scene during the last few years of the monarchy before it was overthrown in 1952. After a brief coexistence with Nasser's regime, the struggle for power between the two ended with the imprisonment of the Muslim Brothers' leaders. They re-emerged in the early 1970s and were allowed to participate in parliamentary elections and those of professional associations.

During 1995, the relationship between the Mubarak regime and the Muslim Brothers became increasingly tense. Some 54 prominent cadres of the Muslim Brothers were accused of illegally re-establishing the group's organisational structure and were tried before a military court. Many of them were sentenced to between three and five years in prison with hard labour.

The security agencies' relative success in curbing the activities of the militants is probably one reason for this escalation of tension. The agencies were thus able to devote more resources to their struggle against the Muslim Brothers. The state may have been applying its own logic by starting with the more militant groups first. Security officials have always officially maintained that members of the militant groups came from the ranks of the Muslim Brothers, believing that the Brotherhood used the *jama'at* simply to exert pressure on the state in pursuit of the same objectives espoused by the militants.

It is also possible that the November 1995 elections were the key to the state's attack on the Brotherhood. The regime was determined to prevent the *Ikhwan*, the most significant opposition force in society, from participating in the election itself as well as from the national dialogue that preceded it. The *Ikhwan* appeals to those in the recently urbanised segments of society who resent rising corruption, high inflation and the widening gap between the haves and the have-nots. Unemployment, which reached 20% and above in parts of upper Egypt, also breeds local conditions conducive to greater Islamist influence. Public opinion polls before the election suggested that the *Ikhwan* could have gained 15% of the vote in free and fair elections. Thus, the pressures on the Muslim Brothers could have been a pre-emptive move to weaken their standing before the parliamentary elections.

To The Polls

Although the Egyptian system of government centres on the presidency, there was some hope that the parliamentary elections in November 1995 would provide an opportunity for genuine popular participation, producing a more assertive legislature in which opposition political forces would have greater representation. The number of candidates competing in these elections was unusually high (about 4,000), the turnout was the highest in recent years (approximately 50%), and 14 opposition parties, and some independents, who had boycotted the 1990 elections, participated.

Notwithstanding these favourable signs, the results indicate greater political exclusion of the opposition. Not a single candidate representing the opposition won in the first round in any district. Moreover, most leaders of opposition parties lost in districts which they had historically held with strong family and local support. The election was also marred by an unprecedented level of violence. According to an estimate by the Centre for Human Rights and Legal Aid in Cairo, 51 people lost their lives and 1,700 were injured in election-related clashes.

When the final election results were declared, the government party, the National Democratic Party (NDP), controlled 317 of a total of 444 elected seats in parliament. Although independent candidates won 113 seats, most of them were members of the NDP who ran as independents only after being excluded by the ruling party from its electoral lists. Since the elections, 99 of these parliamentarians have rejoined the NDP, thus giving it an even greater level of control. This is in sharp contrast to the results of the 1987 elections when opposition and independent candidates, while hardly threatening the NDP's political control, won 102 seats and made parliament a more vigorous forum for debate than ever.

The results of the 1995 elections were appealed in about one-third of the districts. If the judiciary, which has a credible record of independence, invalidates the results, the likelihood of the current parliament completing its term is indeed in doubt. Observers of the elections reported cases of harassment and the detention of political activists and campaign workers and took note of the role of the Ministry of Local Government in influencing the elections in the countryside through state patronage.

The state had also maintained its ban on the formation of new parties. The NDP refused to allow opposition parties to have more than symbolic access to radio and television which remained under government control. The State of Emergency, in effect since 1981, bans marches and mass gatherings and, thus, set serious limits on party activities. Not surprisingly, many observers of the Egyptian political system describe it as a case of 'restricted political pluralism'. Through publicising such restrictions and illegalities, the opposition parties hoped to put serious international, and particularly Western, pressure on the government. But Egypt's strategic importance, and its perceived role in the incomplete Arab–Israeli peace process, have thus far frustrated such opposition hopes.

The most significant aspect of the 1995 elections was the exclusion of the Muslim Brothers from the legitimate political process. Generally speaking, the state adopted a harsher line towards the Islamists after the attempt to assassinate Mubarak. Contrary to earlier expectations that the state might accept an arrangement similar to that worked out in Jordan whereby a moderate wing of the Muslim Brotherhood is permitted to work within the legal framework, the Egyptian regime has insisted that the Muslim Brothers remain a banned organisation. Even the earlier acceptance of Muslim Brotherhood activities in national politics under the umbrella of the Labour Party or the *Wafd,* or of their control over the professional councils, was no longer tolerable. Unsurprisingly, more militant Islamists, such as the *Jihad* group, used these developments to prove the futility of the compromise of working within 'un-Islamic institutions', since the only result appeared to be exclusion from the electoral process and a jail sentence.

An interesting question is whether more state restrictions against the Muslim Brothers, which made working within the boundaries of legality more difficult, would push them towards clandestine activities. In January 1996, Mustafa Mashhour, one of the founders of the secret security branch of the Muslim Brotherhood during the 1940s, succeeded to the position of Supreme Guide on the death of Mohammed Hamid Abu al-Nasr. The few statements he has made since then reflect his interest in legalising a party that represents the Muslim Brothers and struggles through peaceful means to gain power in order to build an Islamic state and society.

Previous efforts of militant confrontation with the state seem to have convinced the Brotherhood leaders of the high price of such a strategy, which only resulted in ever stronger waves of suppression. In 1989, Mashhour made it clear that while commanding power is an Islamic duty, power should not be automatically equated with a resort to violence. As he saw it, the Islamists should learn from past experience, and particularly from the example set by the Prophet Mohammed himself. According to Mashhour, the main lesson to be learned is the crucial need for doctrinal training and organisational cohesion before adopting any confrontational approach in pursuit of political power.

Thus, a protracted stalemate continues to characterise the Egyptian political scene. Islamists are the strongest opposition force. They continue, through an Islamic version of populism, to appeal to segments of society afflicted by the high cost of economic liberalisation. However, Islamist groups are sharply divided over the best strategy to gain or seize state power from the hands of a regime alarmed enough by their combined activism to become more committed than ever before to reducing their influence. The greater the level of violence committed by the militants, the narrower the political space which the regime is likely to allow Islamists in society.

Lessons From the Region?

Like Egypt, other Arab regimes have been facing diverse challenges from social groups that have been unable to bring the regimes down or to destabilise them enough for a third force to seize power. Developments in Algeria, however, suggest that even a regime that relies heavily on the army and security forces to confront radical Islamists and to restore state authority in parts of the country where such authority has virtually withered away, may recognise that in the long run it must legitimise itself politically through elections and political dialogue with opposition forces. President Liamine Zeroual has apparently concluded that although his regime has had considerable success in marching down the military track, any thought that the Islamic opposition can be eradicated in this way is an illusion. Instead, the Algerian regime must seriously explore the political track. In this sense, the presidential election of November 1995 may prove to be a significant landmark in transforming Algeria's political equation.

Presidential elections were opposed by the Islamic Salvation Front (FIS), as well as other secular opposition organisations such as the National Liberation Front (FLN) which signed the Rome Declaration in January 1995. This document pledged support by a united opposition for a dialogue with the regime about how to prepare for free elections and to establish the rule of law which, in effect, would put an end to the military seizure of political power. The more militant Armed Islamic Group (GIA) threatened the candidates, and those willing to vote, with a massive campaign of violence.

Their efforts to prevent a meaningful election failed. One of the most significant aspects of the election was the over 70% turnout by the voters, reflecting the strong feeling of conflict weariness that had developed within the populace. President Zeroual gained 61.3% of the vote, not only because he promised to restore order, but because he committed himself to a programme based on dialogue with all Algerian political forces that renounced violence and declared themselves ready to work within the boundaries of legitimacy. Most importantly, he did not exclude the Islamic Salvation Front from such political dialogue if its leaders met these basic conditions.

A dialogue between the state and the opposition has become possible because the election results, closely monitored by regional and international observers, enhanced Zeroual's political legitimacy. After the elections, many FIS leaders were more willing to meet some of the regime's key demands by explicitly denouncing GIA violence without reservation and by offering a unilateral cease-fire by their armed wing, known as the Islamic Salvation Army (AIS), thus exposing the more radical GIA to extreme military and political vulnerability.

In addition, the post-election government, headed by a former Algerian Ambassador to the United Nations and including representatives of the opposition parties that competed in the presidential elections, was more pluralistic than any of its predecessors. Furthermore, there is an unmistakable rift among the parties that boycotted these elections and that had in the past rejected any dialogue with the regime. Some of them have already opened direct channels of communication with the regime, making it clear that they are no longer bound by the Rome Declaration's stipulations of a united front and collective bargaining. Obstacles to political stability in Algeria are still considerable, but if these four developments continue, political compromise may have a real chance of prevailing over violent confrontation.

In Jordan, King Hussein was forced to move away slightly from his earlier policy of compromise. He was increasingly confronted by a defiant opposition, particularly from Islamist elements in parliament, the press and the professional associations denouncing his pursuit of peace and normalisation with Israel at the expense of Jordan's Arab ties. The regime responded by arresting an outspoken critic, the head of the Engineers Association, Laith Shubailat, despite severe condemnation by professional associations. The King threatened the press and professional syndicates with much stricter regulations in order to prove that Jordanian-style democracy can use its sharp teeth if the critics fail to discipline themselves or cease to behave as a 'constructive opposition'.

In all the embattled regimes in the Arab Middle East, societal challenges in 1995 were clear enough to be identified, but not strong enough to undermine the regimes' grip on power. Often, this grip was maintained through even tighter state restrictions. In the short run, this has brought success for the champions of political order over the need for societal participation. The prospects of long-term political stability for such regimes are not promising, however, unless they can find a way to combine their policy of military confrontation with political compromise.

More Continuity Than Change In The Gulf

During the past year, the situation in the Persian Gulf has continued to display most of the features visible since 1991. While the government of Saddam Hussein continued in power, Iraq remained under the rigorous UN sanctions imposed in 1990. Mistrusted both by its neighbours and by much of the international community, the Iraqi regime did nothing to advance its rehabilitation. It continued to conceal aspects of its weapon-development programmes and acted with characteristic brutality towards its critics. While it entered into negotiations with the UN for a partial lifting of sanctions to pay for food and medicine, it was deeply reluctant to accept all the terms of UN Security Council Resolution (UNSCR) 986, the basis for the UN position.

Iran displayed its customary and unsettling ambivalence towards its Gulf neighbours, alternating professions of friendship with sharp criticisms, while continuing to assert its right to disputed territories in the waters of the Gulf. Fear and suspicion of both Iraqi and Iranian intentions reinforced the determination of the oil-rich Arab states of the Gulf Cooperation Council (GCC) to look to the Western powers, and the United States in particular, for their security.

Reassuring as this may have been for external threats, it did little to help these regimes meet the internal challenges emerging in their domestic politics. Some of these challenges were more serious than others, but their nature and the very different responses they elicited from the GCC governments suggest that under the cover of an apparently frozen regional configuration of power, there are forces at work that may yet provoke changes within and between the Gulf states themselves.

The Pariah State

In the first part of 1995, Iraq was optimistic that the UN sanctions regime would soon be lifted, believing that France and Russia, in particular, would argue in its favour on the UN Security Council. Thus, it initially rejected the proposal for limited oil sales put forward by the Security Council in UNSCR 986. By the end of the year, however, it was clear that there was little momentum in the Security Council for lifting the sanctions. Instead, Iraq entered into discussions with the UN Secretary-General on the terms of UNSCR 986. A number of factors appear to have contributed to this decision.

First, the US remained adamantly opposed to lifting the sanctions. It made clear that, before sanctions could end, Iraq would not only have to comply with the UN's demand for the complete disclosure, and the dismantling, of its non-conventional weapons and missile programmes, but it would also have to fulfil a number of other conditions. These included accounting for the 600 or so Kuwaitis missing since the Iraqi occupation, ending the 'export of terrorism', and suspending the repression of its own people. Some members of the UN Security Council, however, found these conditions insufficient justification for maintaining sanctions. For them, Iraq's full compliance with UN disarmament resolutions would be enough to merit lifting sanctions. The issue was never really tested, however. During the year, Iraq's deception and prevarication over questions relating to disarmament effectively undercut its case at the Security Council.

Iraq had grudgingly cooperated with the United Nations Special Commission on the Disarmament of Iraq (UNSCOM) since 1991, gradually revealing the extent of its chemical, nuclear and missile-development programmes. Nevertheless, the Iraqi authorities maintained throughout that they had no programme to develop biological weapons. Rolf Ekeus, the head of UNSCOM, suspected otherwise and his doubts were reflected in an April 1995 interim report to the Security Council. In it Ekeus charged the Iraqi government with failing to give a complete account of its biological weapons programme. He reiterated these charges in June, stressing that UNSCOM could not report Iraq's full compliance with the disarmament resolutions until it cooperated on the question of biological weapons. As a result of Ekeus' findings, wavering members of the UN Security Council were brought back into line and the sanctions regime was maintained in its entirety.

As with the piecemeal revelation of previous weapons programmes, this setback prompted some partial disclosures by the Iraqi government. In early July, shortly before the UN Security Council was to meet for its bimonthly consideration of the sanctions issue, Iraq admitted for the first time that it had indeed set up a programme to develop and produce biological agents for offensive use – notably botulinum and anthrax. It claimed, however, that these agents had not been converted into weapons and that all had now been destroyed. In a characteristic gesture, Saddam Hussein also threatened that unless the UN sanctions were lifted in their entirety by 31 August 1995, 'there will be no further cooperation with the UN weapons inspectors'. Only Saddam could have been surprised that after this combination of prevarication and truculence the UN Security Council refused to lift the sanctions.

The next move, however, was dramatic and unexpected. In the first week of August, General Hussein Kamel Hasan and his brother Colonel Saddam Kamel, Saddam Hussein's sons-in-law and key figures in his clannish regime, fled with their families across the border to Jordan where they were immediately granted asylum by King Hussein. The significance of this was twofold. First, it demonstrated that even Saddam Hussein's apparently resilient family-based system of power was under strain. Second, Hussein Kamel Hasan, as Minister of Military Industry, had been in charge of building up Iraq's non-conventional weapons programme.

An extraordinary spectacle followed. Both Hussein Kamel on the one hand, and the Iraqi government on the other, hastened to divulge details of Iraq's biological and other weapons programmes to UNSCOM. Hussein Kamel was evidently trying to ingratiate himself with his new patrons and to further blacken his father-in-law's name, while the Iraqi government hoped to blame all the subterfuges of the preceding years on Hussein Kamel.

As a result, Iraq handed over some 700,000 pages of documents to UNSCOM and made further disclosures to the International Atomic Energy Agency (IAEA) about its nuclear programmes. But Iraq's hopes were dashed again. In his interim report to the UN Security Council in October, Ekeus placed the blame squarely on the Iraqi government for having misled the inspectors on the scope of its weapons programmes. Although UNSCOM had not had the opportunity to read all the documents, it was nevertheless clear that Iraq's biological weapons programme was far more extensive and ambitious than had first been thought. And Iraq's claims that it had destroyed its weapons of mass destruction were far from being adequately verified.

In Ekeus' second bi-monthly report to the Security Council in December, he indicted Iraq for continuing to withhold information from UNSCOM and for misleading the Security Council, particularly over Iraq's missile and biological weapons programmes. The UN feared that Iraq was still concealing both missiles and warheads armed with biological weapons.

Tightening The Grip

In January 1996, UNSCOM inspectors recovered a consignment of recently hidden gyroscopes suitable for long-range missiles from the waters of the Tigris, close to Baghdad. This seemed to indicate that Iraq's missile-development programme was indeed continuing. The discovery came in the wake of a seizure of similar gyroscopes by the Jordanian authorities in a shipment bound for Iraq in December 1995, and coincided with the seizure of a hidden consignment of chemicals in transit through Jordan in January 1996. Because of these developments, Iraq knew that UN sanctions were not about to be lifted, and this may have led it to start open discussions on the limited sale of oil.

Jordan's involvement in these events was also significant. In recent years, King Hussein had been trying to move away from the relatively close relationship his country had developed with Iraq. During 1995, this policy, personified by the Minister of Foreign Affairs, Abd al-Karim Kabariti, subsequently Jordan's Prime Minister, became more focused. Jordan's rather loose enforcement of the sanctions regime against Iraq was tightened.

The defection of Hussein Kamel Hasan and his brother to Jordan in August 1995 encouraged King Hussein to advocate a complete change of regime in Iraq. After granting Hussein Kamel and his entourage asylum in Jordan, the King seemed briefly to promote Kamel as an alternative Iraqi leader. Although this came to nothing, it shed some light on the King's thinking. Underlying his action was the belief that a change in Iraq's regime would only occur if the Sunni Arab clans of the Iraqi officer corps were persuaded to act against Saddam Hussein.

At the same time, King Hussein strengthened enforcement of the restrictions on trade with Iraq, ordering that goods shipped to Iraq in transit through Jordan be more rigorously monitored. This severely reduced the opportunities available to much of the Iraqi elite to use their connections with Jordan to evade the worst effects of the sanctions. It also revealed Iraq's continuing attempts to maintain and develop its weapons programmes using Jordan as a transit point for importing appropriate technology.

These moves to cut the Iraqi elite's lifeline had a considerable impact within Iraq. One of the reasons for Saddam Hussein's political survival in the aftermath of the 1991 Gulf War and under the punishing sanctions regime had been his ability to ensure that goods and economic resources continued to flow to the infamous 'five hundred thousand'. This is roughly the number of the elite and their dependants who constitute Saddam Hussein's support base. Hitherto, Saddam had used the vast sums siphoned off from Iraq's oil revenues over the past 25 years to maintain the loyalty of these people. In addition, the elite has developed various sanction-breaking ploys to maintain their privileges, whether these be oil sales through Iran, the shipment of goods and oil through Kurdistan to Turkey, or the thriving trade through Jordan.

Figure 1 *UN Security Council Resolution 986*

- Sale of Iraqi oil not to exceed $1,000 million every 90 days. Continuation of sale to be reviewed after 180 days.
- Purchase and transport of oil to be monitored by the UN.
- Oil to be transported through Turkey's Kirkuk-Yumurtalik pipeline.
- Proceeds of sale to be placed in account established by the UN Secretary-General.
- Of the $1,000m:
  - *a* $300m to go on war reparations;
  - *b* $650–680m to go on medicine, health supplies, food and other supplies for civilian needs, of which $130–150m must be given to the UN Inter-Agency Humanitarian Programme operating in the Kurdish north; and
  - *c* The rest to cover the cost of the UN operation and supervision of the oil sale and repairs to the pipeline.
- Plan for an equitable distribution of humanitarian supplies to be submitted to and approved by the UN Secretary-General.

After five years of depletion it seems that even Iraq's hidden reserves are running low. The restrictions imposed by Jordan on trade dealt a further blow to the regime's ability to retain the confidence of the elite. These considerations may have helped to push Iraq into negotiations with the UN Secretary-General over UNSCR 986. The Resolution, based on a proposal put forward by the US and the UK, the states most strongly opposed to lifting sanctions, had been passed unanimously by the UN

Security Council in April 1995. Although these countries thought it a generous offer, the Iraqi government found its terms far from ideal.

Not only does Iraq regard the terms of the Resolution as an infringement of its sovereign right to dispose of its national income as it chooses, but the Resolution also singles out the Iraqi Kurds for special treatment, possibly indicating a future special status outside the sovereign jurisdiction of the Iraqi state. Furthermore, the availability of such sums to finance importing food and medicine permitted by the UN trade restrictions would blunt the propaganda impact of the campaign conducted by the Iraqi government during the past five years. This has emphasised the plight of the Iraqi people as a way of pressuring for sanctions to be lifted in their entirety. These considerations had caused the Iraqi government to reject a similar proposal in 1993, and formally to reject UNSCR 986 in April 1995.

However, the combination of pressures on the resources available to the Iraqi regime and the realisation that the sanctions as a whole would not be lifted in the near future, seems to have caused the Iraqi government to change its mind. It may also have changed its strategy. Iraq may now believe that engaging with the UNSC through Resolution 986 and the return of Iraq as an oil producer will begin to normalise relations between Iraq and the rest of the world, thereby creating the necessary momentum at the UN to lift the sanctions.

But that outcome remains as uncertain as ever. The fact that UNSCR 986 was proposed by the US and the UK would suggest that they intended it to lessen rather than to increase the pressure for sanctions to be lifted. This consideration cannot have been lost on the Iraqi government and, quite apart from the questions of sovereignty, seems to have bolstered Iraq's reluctance to accept the Resolution. The negotiations which began in early February 1996 between the Iraqi delegation and the UN ended inconclusively a fortnight later. A second round of talks in mid-March in New York brought no agreement beyond establishing 8 April as the date for a third meeting.

Domestically, the resources liberated by such a deal would help Saddam Hussein to sustain his patronage network. He is also determined to convince the rest of the world that he and his regime are here to stay for the foreseeable future. To this end, he organised a presidential 'election' in October 1995 in which he was the sole candidate and in which he ensured that he received 99.96% of the votes. This was intended to send the message to the rest of the world that, however unpalatable some may find it, they would have to come to terms with him again. This argument seems to have succeeded with some members of the UN Security Council, notably France and Russia. However, it has yet to convince the US administration, which advocates the strategy of 'dual containment' as a way of dealing with potential security threats in the Persian Gulf.

The Squabbling Kurds

It is certainly true that during the past year the Iraqis themselves seemed no nearer to overthrowing their President. The external opposition to Saddam Hussein continued to proliferate and to fragment. An attempt to unite this opposition under the umbrella of the Iraqi National Congress (INC) seemed to unravel as one group after another chose to dissociate themselves from that body during the year. In December 1995, King Hussein's new interest in Iraqi opposition politics led him to call for a conference intended to be held in Amman in January 1996 to discuss 'post-Saddam Iraq', based on his notion of a federal Iraq. This proposal received a very mixed response from much of the Iraqi opposition, a number of whom chose to believe that King Hussein was advocating the permanent division of the country along ethnic and sectarian lines.

One of the main problems plaguing the credibility of the INC, and of the Iraqi opposition more generally, was the continued bitter conflict between two of its main constituent parties: the Kurdish Democratic Party (KDP); and the Patriotic Union of Kurdistan (PUK). Open fighting between these two groups since 1994 has cost over two thousand lives, and more were lost when fighting again flared up in July 1995. The seriousness of the situation, as well as the apparent willingness of the Iranian government to help mediate the conflict, led the United States to attempt to broker an accord between the two rival parties. Under its auspices, the United States brought representatives of the KDP and PUK together for two meetings in Ireland in August and September. No agreement was made, however, and the conflict continued sporadically.

Turmoil in Iraqi Kurdistan caused a number of interested parties to intervene during the past year. Most dramatically, Turkey launched a massive invasion in March 1995, involving 35,000 troops, that lasted for six weeks. The Turkish forces were attempting to pursue and destroy units of their own Kurdish resistance movement, the Kurdish Workers' Party. As with previous operations, the results were inconclusive, but they did heighten the impression that the Kurdish zone has no effective administration. Despite this, international outcry at the scale of the Turkish invasion appears to have discouraged Ankara from its initial idea of establishing a buffer zone inside Iraq along the Turkish frontier. The Turkish forces withdrew in April 1995, but, almost inevitably, launched another military offensive into northern Iraq in July.

Iran also used the disarray of the Kurdish zone to play a role. Links already existed between Tehran and the Islamic Movement of Kurdistan (an ally of the KDP). Increasingly, however, Iran saw some benefit in proposing itself as an even-handed intermediary between the rival Kurdish groups. It therefore encouraged visits by KDP and PUK representatives to Tehran, as well as trying to encourage the warring sides to disengage in some areas. After the visible failure of the US-sponsored Dublin 'reconciliation talks' in September, it was significant that the leaders of the KDP and the PUK returned to Tehran. If an effort towards reconciliation between the KDP and PUK was attempted there, however, Iran seems to have had no more success than others.

Signs of Unrest

Saddam Hussein could only take comfort from the confusion in the Kurdish zone and was evidently gratified to hear voices raised in Kurdistan advocating a reconciliation between the Kurdish parties and the Baghdad government. More pertinently for the future of Saddam Hussein's regime itself, in the first half of 1995, as in previous years, there were confused reports of abortive coup attempts within the Iraqi armed forces, usually involving a mere handful of officers. The only significance in these developments was that the conspiracies were apparently hatched among clans of the Sunni Arabs who had hitherto been both the mainstay of the Iraqi officer corps and of Saddam Hussein's regime. The aftermath of one such alleged conspiracy led to riots and some violence among the Dulaim tribe of Anbar province in May and June, but this was suppressed with little difficulty by forces loyal to the regime.

The defection in August of Saddam Hussein's sons-in-law appears to have been the result of clan rivalries within the regime and not an attempt to overthrow the government. In particular, it seems to have been provoked by the growing power and influence of Saddam Hussein's two sons, Uday and Qusay, which has caused resentment among other members of the family, including Saddam Hussein's half-brothers. Some of Uday's privileges were removed in the immediate aftermath of the August events, but by the end of the year he was as visible and as powerful as ever.

The fact that he and his brother are disliked within the wider family, in the officer corps and among the Sunni Arab elite more generally, is potentially dangerous for Saddam Hussein should he persist in promoting them to the exclusion of others. However, Saddam has demonstrated his skill and ruthlessness at playing the game of clan and patronage politics. In February 1996, he persuaded his two sons-in-law to return to Baghdad, where they were promptly murdered, reportedly by other members of their clan which is also that of Saddam Hussein himself. The violence of these events so close to the heart of power may be an ominous development for the resilience of his regime. For the moment, although Saddam Hussein is by no means immune either to the assassin's bullet or to a military *coup d'état*, he seems to be as entrenched in power as ever five years after the defeat of his forces in Kuwait.

The Arab Gulf States

There were some new developments in the Gulf during 1995, alongside a continuation of trends in evidence since at least the end of the 1991 Gulf War. The United States continues to play the key role in defence of the Gulf states and their security plans, as seen in the March 1995 agreement between Qatar and the US over prepositioning US military equipment on Qatari soil. Different perceptions of the potential threat from Iraq continue to divide members of the GCC, with Kuwait and Saudi Arabia maintaining a hardline stance against the rehabilitation of Iraq until it complies fully with all UN Resolutions.

The attitude of these two states set the tone of the GCC summit resolution in December 1995 which harshly condemned Iraq, despite the fact that, during the preceding year, Qatar and Oman, but also the United Arab Emirates, had advanced suggestions for softening the approach to sanctions. And relations with Iran were as chequered as ever. Unspoken mistrust of Iranian intentions was periodically fanned into open denunciation over the unresolved dispute over the Gulf islands of Abu Musa and the Tunbs. The GCC continued to support the UAE's claim, while Iran maintained its insistence that it was sovereign. Despite the rhetoric, this is still a war of words and, given the capabilities of the two sides, seems likely to remain so.

Shaky Regimes

Bahraini and Saudi accusations against Iran over the continuing civil unrest in Bahrain sharpened considerably during 1995. The trouble began in December 1994 when the Bahraini authorities arrested a number of its citizens who were intending to present a petition to the ruler demanding the restoration of the Constitution and the re-convening of the representative assembly; both had been suspended in 1975. These arrests led to protests and rioting, largely in the Shi'i areas of the island, which were suppressed with considerable violence by the security forces.

In March, rioting re-erupted and was again suppressed, leaving a number of dead on both sides. Thereafter, some Bahraini prisoners were released, but many remained in detention, and the government did little to begin dialogue with the opposition, rejecting outright demands for the restoration of the Constitution. Instead, the Bahraini government sought to blame the unrest on Iranian subversion and continued to tighten security through purges and arrests.

Continued unrest and discontent at this situation among the Shi'i (comprising over half the population) led to further open rioting in January 1996, to which the authorities responded with even more severe repression. Although concentrated among the more deprived, Shi'i population, opposition to the government's authoritarian methods is by no means confined to this group, as increasing numbers of arrests among Sunni Bahrainis prove. Nor can the unrest be said to be inspired principally by Iran. Rather, Bahrain is undergoing what happens when a ruling family seeks to maintain its exclusive hold on power in the face of a population increasingly unwilling to accept the old rationales for such a system of government.

Although the same conditions do not apply equally throughout all the Gulf states, an awareness of the need to respond to new pressures has shaped the rulers' policies during the past year, albeit in very different ways. In Kuwait, the franchise for the coming parliamentary elections late in 1996 has been greatly extended to encompass a large number of people hitherto excluded, thereby reinforcing the legitimacy of the national assembly. By contrast, in Oman, restrictions have been tightened, leading to purges and arrests of adherents of various Islamist political organisations.

In Saudi Arabia, similar measures have been implemented to restrict the influence of Islamist organisations, such as the Committee for the

Defence of Shari'a Rights. This has not, however, prevented new manifestations of anti-regime feeling. In November 1995, a bomb exploded at the Saudi National Guard base in Riyadh which housed US military advisers assisting in training the Guard. Seven people were killed, including four Americans.

This was an unprecedented attack. It drew attention to underground opposition to the Saudi regime, as well as to the strong security relationship between the US and the Saudi government. Responsibility for the explosion has yet to be established, and it was not followed by any further acts of violence. Yet the nature of the targets and the act itself had wide reverberations in Saudi society.

No less dramatic was the stroke suffered by King Fahd in December. It left him too ill to carry out his duties and in January 1996 he handed over authority for the day-to-day running of the kingdom to Crown Prince Abdullah. This transfer of power was of short duration: the King recovered and resumed his full powers in late February. Despite reports of differences between Prince Abdullah and other senior Princes, as well as with the King, on some matters, there do not appear to be any serious differences on fundamentals. Equally, although there are questions about the line of succession once the new generation becomes eligible, for the moment there is wide agreement on succession and the consensual mechanisms of the Saudi ruling family seem to be working appropriately.

Cracks in the GCC Wall

The most dramatic event of the year in the Gulf region took place in Qatar in June 1995. Crown Prince Sheikh Hamid bin Khalifa al-Thani seized the opportunity of his father's absence from the country to proclaim himself ruler of the state. The significance of this palace coup was twofold. On the one hand, Sheikh Hamid is thought to be better able to guide Qatar and the interests of the ruling family through the uncertain years ahead, whether by sound economic management, or by opening up to some form of representative political life, as he has hinted.

On the other hand, he has long been associated with Qatar's distinctive and independent (chiefly of Saudi Arabia) foreign policy: he has succeeded in the difficult task of befriending Iraq, Iran, Israel and the United States. At the same time, however, Qatar's relations with Bahrain and Saudi Arabia have become increasingly embittered, principally over bilateral territorial disputes. These conflicts have given rise to unusual public displays of antipathy, disrupting the veneer of studiously maintained calm in inter-state relations in the GCC. One recent indication of such tensions was Qatar's criticism of other GCC members for failing to denounce an attempted counter-coup launched by its former ruler in mid-February 1996.

In December 1995, Sheikh Hamid walked out of the GCC summit meeting in Muscat, Oman, apparently in protest at Saudi dictation of the final communiqué and at the selection of a Saudi as the next Secretary-General of the organisation. Moreover, Bahrain and Saudi Arabia made a point of receiving with full honours the deposed ruler of Qatar who was touring the GCC states in an attempt to muster support for his re-instatement. The fact that the deposed ruler had been able to lay his hands on an estimated $5bn of Qatari funds gave him a strong financial base from which to promote his attempt to recover power. Although he was also received in Kuwait and the UAE, it was Bahrain that received the full brunt of Qatar's criticism. In retaliation, Qatar screened a long television interview in January 1996 with two of Bahrain's most prominent exiled dissidents, predictably outraging the Bahraini government.

Such disputes have, of course, simmered under the surface of GCC relations for many years. The difference now is the increasingly public nature of these quarrels. This may indicate that not all the states are still prepared to play by the same rules. This new willingness to express disagreement probably reflects a combination of domestic developments and the states' direct security relationships with the West, and the US in particular, which have made them see less value in the GCC and the conformity with Saudi Arabia's wishes which membership entails.

Behind The Veil

Little seemed to have changed over 1995 in the regional balance of power in the Persian Gulf, or in the capabilities of the states concerned. Such apparent stasis may be deceptive, however. Internally, processes of challenge, reaction and adaptation indicate that the domestic politics in the Gulf littoral are far from dormant. In Iraq, the inner circles of the regime, hitherto solid in their support of Saddam Hussein, have shown themselves to be susceptible to fragmentation and internecine feud. In Iran, the campaign leading to elections to the new *Majlis* in April 1996 highlighted for many the division in Iranian politics between those who adhere to an unreconstructed ideological vision of Iran's future direction and those who seek a more conventional place for Iran in the world. In the Arab states of the GCC, no radical change occurred, but the forces pressing for change of varying kinds became more visible. This both provoked a variety of responses from the rulers of these states, and, in some cases, complicated relations between them. These trends may well shift the foundations on which the security configuration in the Gulf has been based for the past five or so years.

Asia

Alarm bells tinkled and then rang loudly in Asia in 1995–96. First, China seized new territory in the South China Sea, and then effectively closed major air and shipping lanes in the Taiwan Strait as it tried, with massive military manoeuvres, to intimidate Taiwan. Even Japan and South Korea revived an old territorial dispute. If any still dreamt that economic inter-dependence in East Asia was sufficient for peace and stability, they awoke from their reverie to find that only a robust demonstration of US naval power could keep a stable balance. Where were the much-vaunted Asia-Pacific Economic Cooperation (APEC), the Association of South-East Asian Nations (ASEAN) or the ASEAN Regional Forum (ARF) when it was time to keep the peace?

This was also a year of economic anxiety throughout the region. Stock markets in emerging countries slowed, Japan's economy continued to stagnate, and China and Vietnam lost momentum for their economic re-forms. By contrast, India's economic reforms seemed more sprightly, but electoral uncertainty loomed. Concern about succession politics in China, North Korea, Indonesia and even Singapore was accompanied by political scandals in South Korea and leadership struggles in Japan. Somehow, Asia appeared a less miraculous, and far more vulnerable, region in 1995–96 than had previously been promised.

China's Edgy Vigil

The long wait for China's patriarch, Deng Xiaoping, to die has produced a scarcely concealed and potentially costly succession struggle. Deng has reportedly expressed a desire to live long enough to see Hong Kong handed back to China on 1 July 1997. Although still alive, he is apparently often virtually comatose. As a result, his country is suffering from a lack of clear leadership. Another year of waiting risks weakening Deng's greatest achievement – China's economic reforms. The need for each of China's putative leaders to be more ruthless than the next in dealing with the outside world threatens once again to isolate the country just when it would most benefit from regional, and global, economic and political integration. Deng has worked heroically during his long life to make contemporary China a more prosperous, modern nation; yet his tenacious grip on life at this critical junction may actually be damaging his country.

What Direction For the Economy?

China's economy is a paradox. No other major country would consider a 1.6% decline in the growth rate (the target was 3%) a success. But after several years of accelerated economic growth, China's leaders and foreign advisers have been attempting to decelerate the Chinese economy. The official gross domestic product (GDP) growth rate in 1995 was 10.2%, down from 11.8% the previous year. While Chinese statistics are best seen as notional rather than reliable, the trend is clear – after three years of trying, the Chinese government has been able to slow the dynamos. Inflation, at least according to official data, also fell from 22% to 17%.

Figure 1 *Real GDP Growth*

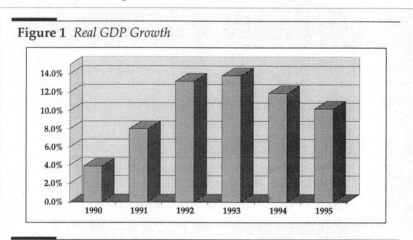

Figure 2 *Inflation*

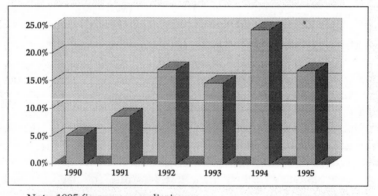

Note: 1995 figures are preliminary.
Sources: International Monetary Fund, Chinese government.

These headline figures, encouraging as they are, hide more serious problems. The growth and inflation targets were set three years ago and the delay in reaching them says much about the only partial extent to which the central government controls China's economy. After years of missing targets, the Chinese government resorted to much blunter, less market-oriented methods in 1995, thus starkly revealing the limits of economic reform and the challenges still to be met.

A key weapon in the war against inflation was Beijing's crude tightening of price controls on the few basic commodities still under central control. Not surprisingly, and as noted by the State Information Center in December 1995, lower grain prices dampened 'the enthusiasm of farmers' and increased the risk of grain shortages and unrest in the countryside. The 250 million city dwellers want lower prices, but the 950 million rural population want them higher. An Organisation for Economic Cooperation and Development (OECD) report in November 1995 suggested that unrealistic grain prices may cause China's grain imports to reach 50m tonnes a year by 2000.

At the heart of the economic challenge lies reforming state-owned industries. Little progress in this respect was made in 1995. Despite more than 10% growth in the overall economy, the losses incurred by these state industries increased by some 20% in 1995, and 40% were still operating in deficit. They continued to be a major drain on central government finances, leading the government to slash spending on vital fixed-asset investments. In short, the financial black hole at the heart of the Chinese economy created by these state industries was as serious a problem as ever. Without reforming these industries, the results of broader economic reform risked evaporating into thin air. The failure to collect more than 20% of the new Value Added Tax, and the inability to pay exporters their $6 billion in tax rebates, suggest that reforms introduced into the tax system did not in fact fill the central government coffers. Tax reform was intended to help solve the problem posed by state industries. In 1994 (the latest available figures) central government revenue was 5.1% of GDP (down from 6.8% in 1992), compared to 14% in Japan, 20% in the US and 35% in the UK.

The publication of China's Ninth Five-Year Plan, to begin on 1 January 1996 – in any case an anomaly in a country with more than 50% of its GDP in non-state hands – was thus greeted with more amusement than awe. The average annual growth rate was projected to be 9% with inflation running at less than 10%. The fate of the Eighth Five-Year Plan suggests that these figures are even less useful than the OECD's five-year projections for Western economies.

China's incomplete adoption of market reforms, and the underlying desire of officials to buck, more than to buckle under, the market, was also evident in China's response to pressures to open its markets to foreign trade and investment. In 1994, China believed that it could become a founder member of the successor to the General Agreement on Tariffs and

Trade (GATT), the World Trade Organisation (WTO). In 1995, it not only discovered that one-time trade concessions were not sufficient for membership, but that China would be asked to make continuous concessions to the common operating procedures of the global market economy. China's leaders were clearly in two minds about whether to accept such constraints inherent in genuine economic interdependence, or whether to find some way of liberalising just enough, but without surrendering real sovereignty.

Figure 3 *Trade Balance*

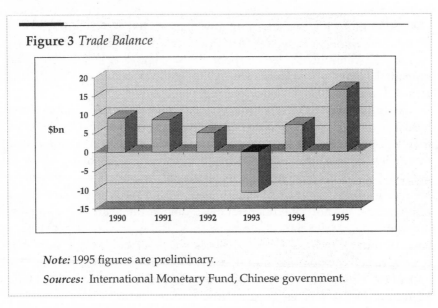

Note: 1995 figures are preliminary.

Sources: International Monetary Fund, Chinese government.

At the APEC summit on 19 November 1995, China stole the headlines with a promise to cut 4,000 tariffs by 30%. When foreign journalists and governments asked for details, it became clear that Beijing had not yet decided which tariffs to cut, believing that expressing good intentions would suffice. With an average tariff rate of 36% in 1995 (the average for developing states is 15%), it was clear that China had a long way to go before it could meet WTO standards. Pirated software sales in China stood at 98% of total software sales, compared to Brazil's 77% (the average for developing states). China was also far from meeting international standards on intellectual property rights, and in December 1995 Chinese CD pirates took out contracts on the lives of international observers from the CD industry and forced the closure of their office in southern China. In January 1996, China announced curbs on foreigners reporting Chinese economic data – another major step backwards from the transparency in economic relations vital for membership of the WTO and for continued foreign confidence in the Chinese market.

The effort to curb the exchange of economic information, including restrictions on access to the Internet announced in February 1996, suggested that China was drawing back from key features of Deng Xiaoping's strategy of opening to the global economy. Indeed, in early 1996 there were reports that Deng's would-be successors were making barely veiled criticisms of his 1993–94 efforts to speed up economic reform and of his broader strategy of rapid economic growth. Clearly the confusion in economic policy was part of a wider political struggle for the fate of China.

A Political System Adrift

The stagnation in economic policy went hand in hand with political stagnation. The surface calm was trumpeted as a sign of President Jiang Zemin's firm and ever-increasing grip on power, but a more accurate description would be the calm before the coming storm.

The fifth plenum of the Communist Party's 14th Central Committee, held on 25–28 September 1995, was notable for its low-key communiqué. It approved the Ninth Five-Year Plan and tried to show that a post-Deng leadership was already in charge. Yet the then Mayor of Beijing, Chen Xitong, was formally dismissed from the Politburo, notionally on charges of corruption, but more probably because of struggles for power. It is doubtful that corruption at the highest levels was confined to Chen, but his removal served the interests of various contenders for the post-Deng leadership who wished to appear tough on crime.

All was not as it seemed in Beijing's leadership politics. The appointment of Zhang Wannian and Chi Haotian as Vice-Chairmen of the Central Military Commission was interpreted by many as a sign of Jiang Zemin's strengthening hold over the armed forces. But the evidence was flimsy and unlikely to reflect how a fierce post-Deng power struggle would be resolved. Beijing's apparent decision to reduce the armed forces by 500,000 from the current level of 3.1 million in the coming years was also seen as a sign of Jiang's strong hold on power. More likely explanations included the strain on government spending and decreasing concern over land-based threats to Chinese security. If Jiang was really so obviously in charge in Beijing, at a minimum more progress would have been made on economic reform.

The political stagnation left plenty of room for other forces to develop their own power bases. Crime rates continued to soar and Chinese officials spoke publicly about the problem. In major international surveys of corruption in 1995, China was rated among the top three corrupt nations. Foreign businesses complained about the rising price of protection money and the increasing risk of murder and kidnapping.

Growing drug addiction and related crime also made headlines, even in China's official press. The number of drug-related crimes in 1994 alone was 30% higher than in the entire 1985–90 period. Foreign governments, and especially US State Department officials, complained about China's inability to control the flow of drugs through its territory. China also emerged as a major transhipment route for South-east Asian drugs, and yet Beijing refused to share counter-narcotics intelligence with foreign governments.

Smuggling was also a growing problem, especially along the coast. Even China's state-run oil company, SINOPEC, reported losses of at least $300m a year because of illegal oil smuggling into southern China. Tax revenues were also lost in the process. Chinese pirates hijacked ships up and down the coast and the International Maritime Bureau suggested that the inability of the Chinese authorities to control the problem (some Chinese officials were complicit in such operations) was the main factor behind the rise in piracy in East Asian waters in the 1990s. Foreign firms with major investments in China complained that smuggling was so severe that their own products were coming in to China more cheaply than their factories there could sell to the Chinese market.

Many of these social problems were related to the broader trend of decentralising powers once held by central government. The fact that the central government was increasingly willing to publicise these problems suggested just how much it recognised that its control over key parts of the economy and society was waning. Mass migration from countryside to city continued to be a major part of the process. Some Chinese officials said that up to one-fifth of the population of major cities was made up of recent migrants, with all the attendant social problems of crime and poor living conditions. These trends are common in other parts of the rapidly developing world, but in China they are developing especially swiftly, and on an especially vast scale.

Foreign Policy: The Jitters of Waiting

China's uneasy combination of rapid growth but unstable internal system has fuelled an equally uneasy foreign policy that combines robust nationalism with the pressures of international interdependence. With a leadership struggle looming, it is not surprising that nationalism seems to be winning most struggles. No Chinese leader wants to appear weak when dealing with the outside world.

The swagger of these leaders was seen vividly in the most recent crisis in the Taiwan Strait that began in May 1995. In the run-up to parliamentary elections in Taiwan in December 1995, and presidential elections the following March, Taiwanese politicians began stressing their commitment to greater national self-respect, and in some cases even independence, for Taiwan. President Lee Teng-hui sought, and eventually (despite White House reluctance) obtained, a visa for a private visit to Cornell University,

New York, to receive an honorary degree. China protested at what it considered to be tacit US support for Taiwan's bid to enlarge its political identity, believing, probably with good reason, that it was losing the struggle to force Taiwan to reunify with the mainland. Despite growing cross-Strait trade, Taiwan was realising that it did not have to accept unification on Chinese terms.

Having failed to persuade the Americans to rescind the visa, China reduced its official contacts with the United States and began 'test firing' already well-tested missiles into the waters near Taiwan. China mobilised forces from many different divisions and military regions in a series of 'exercises' to warn the Taiwanese of the risks of resisting Chinese wishes. Foreigners also became the target of China's 'idiom of military action', especially when it unilaterally closed air and sea-lanes near Taiwan while the People's Liberation Army (PLA) played war games.

The Taiwanese government and people did not appear intimidated. In 1996, the government continued to seek visas from the United States for transit stops for low-level officials, and the United States quietly complied. The December 1995 parliamentary elections produced marginal gains for the pro-independence Democratic Progressive Party (DPP), and for a rump of the ruling Kuomingtan (KMT) committed to less independence. The results indicated a re-ordered, more sophisticated, multi-party, coalition-politics electoral system. In the presidential elections on 23 March, held in the midst of China's massive naval manoeuvres, President Lee obtained 54% of the vote. The DPP vote dropped to 21%, and the rump KMT candidate most congenial to Beijing finished third with 14% of the vote. President Lee's success, in the first free election for a chief executive in 5,000 years of Chinese history – and in the teeth of Beijing's opposition – was evidence of the growing maturity of Taiwan's civil society. As this society becomes more robust, the demands for *de facto* independence will grow, as will the risk of further crises with China.

The Taiwan Strait crisis was the most important contributory factor in the sharp deterioration of relations between the United States and China, although trade disputes also soured the atmosphere. China's trade surplus with the United States continued to climb steeply, while Chinese firms blatantly broke already agreed intellectual property right accords. Trade disputes and continuing Taiwanese requests for transit visas suggested that the downturn in Sino-US relations was not temporary. But the Chinese and US foreign ministers met at the ASEAN Regional Forum in Brunei on 1 August 1995, deflecting further deterioration, at least in the short term. When China resumed its military threats against Taiwan in early 1996, however, Sino-US relations worsened. In January 1996, the US made public that it had sent its *Nimitz* aircraft carrier through the Taiwan

Strait the previous month. Chinese officials then told former US Under-Secretary of Defense, Chas Freeman, that they doubted the United States would defend Taiwan because it would be reluctant to 'trade Los Angeles for Taipei'. Such a crude nuclear threat to the United States, and growing Chinese nationalism, suggested that normal Sino-US relations were heading for more conflict than cooperation. To avoid this, US Secretary of Defense William Perry stated in mid-February 1996 that all countries involved should calm the atmosphere; at the same time, he made clear that the United States would live up to its treaty commitment to protect Taiwan from any attempt to change its status by force.

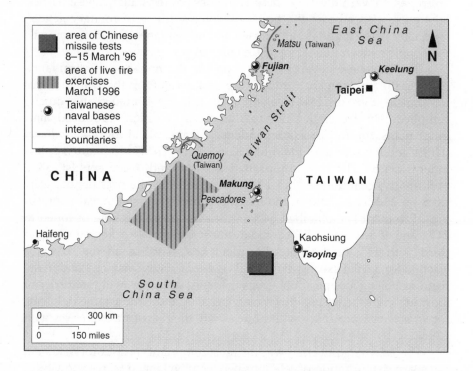

When, before and after the Taiwanese election, China carried out three military exercises in February and March 1996 – with the implicit threat that it would use force to reunite Taiwan with the mainland – the United States abandoned its previous policy of 'strategic ambiguity' that had left some in doubt as to whether it would actually defend Taiwan. The US moved two aircraft carrier battle groups close to Taiwan and warned China not to use force to settle the dispute. This was the largest US military deployment in Asia since the end of the Vietnam War, and was a robust demonstration that the United States had the will to defend the regional balance of power. Although many Asian states declined to support the US action, they also declined to criticise it. Japan and even Singapore suggested that China had erred in threatening Taiwan and offered barely veiled support for the demonstration of US will and power. China may well have made a major strategic mistake, for by loudly indicating that it would not be constrained by its interest in economic interdependence, and by only bowing to US military might, China demonstrated the risk it posed to regional security, and the need for the region to trust in a countervailing military power, not just fine words or economic self-interest.

At the heart of the Sino-US row was China's unwillingness to accept the strictures of interdependence with the outside world. This certainly explained why China did not feel obliged to make far-reaching trade concessions to the United States and the WTO, nor to concede much to its ASEAN neighbours concerned about Chinese activity in the South China Sea. At the August 1995 ARF meeting in Brunei, China tried to portray its acceptance of the 1994 United Nations Convention on the Law of the Sea (UNCLOS) – which it has signed but not ratified – as a concession to ASEAN. In practice, China continued to claim uncontested sovereignty over disputed territory and saw the application of UNCLOS as a way of ensuring international recognition of China's rights over the region's waters and resources.

China also accepted the need to publish a White Paper on defence policy. When it essentially re-published official news-agency material on 16 November 1995 and called it a White Paper on arms control it became clear that for China confidence-building measures need not restrict its sovereignty nor require real transparency. To be fair to China, ASEAN states also put little faith in such cosmetic measures. Indonesia, the largest ASEAN state, signed a defence accord with Australia in December 1995, in effect telling China that it trusted a multipolar balance of power in East Asia, rather than arms control, as a way of managing China. Increased anti-Chinese rioting in Indonesia in 1995 and 1996 also suggested that the role of ethnic Chinese in South-east Asia was far less settled than the ARF liked to pretend. At the end of 1995, concerns about China were nearer the top of the ASEAN agenda than they had been at the beginning of the year.

China's uncompromising attitude over sovereignty also provoked other larger neighbours to trust in the balance of power rather than diplomatic rhetoric. India and the United States carried out joint naval exercises in the Indian Ocean in early 1996. Japan openly criticised China's continuing programme of nuclear-weapon tests on 15 May 1995, and for the first time cut grant aid (albeit symbolically) to China. In August 1995, Japanese aircraft scrambled to intercept Chinese warplanes in Japanese airspace. In November 1995, Japan formally revised its defence policy, putting less emphasis on the Russian threat and expressing more concern about China. As Japan's trade deficit with China continued to grow, Tokyo followed the United States and the European Union in demanding tough conditions for Chinese entry into the WTO. In February 1996, Japan's Diet approved extending its 200-mile exclusive economic zone at sea to include the Senkaku islands disputed with China. Clearly, Tokyo was not prepared to be treated like the ASEAN states.

The other great power in East Asia, Russia, was less willing to criticise China. Russia reportedly agreed to sell more SU-27s to China, and even more importantly, agreed to transfer the technology for making such aircraft. Russia had experienced first-hand what it interpreted as Western coercion and so had some reason to keep China friendly and the West uneasy. China and Russia apparently made some progress towards resolving their border disputes, although Russia had grown increasingly concerned about the massive influx of Chinese across the frontier and re-imposed stricter border controls in 1995. Russian President Boris Yeltsin's heart trouble delayed his visit to Beijing in late 1995, thus postponing the formal signing of the agreements on confidence-building measures along the frontier.

In 1995, China did manage to reach what was likely to be its last major agreement with the UK concerning Hong Kong. In June, it was accepted that there would only be one independent foreign judge on the Hong Kong Court of Final Appeal (CFA) when the colony is returned to China in 1997. The UK had essentially accepted China's terms. Elections held in September 1994 and March 1995 under Governor Chris Patten's plan produced a massive anti-China vote, and prompted Beijing to reiterate in even stronger terms its decision to replace the Legislative Council (LEGCO) when China regains sovereignty. In early 1996, Beijing set up a Preparatory Committee or 'second kitchen' – in essence a rival body that would be treated with increasing legitimacy as 1997 approached. The Committee, which had no representation from the Democrats who won the last free elections in Hong Kong, would ratify China's appointment of the new Chief Executive to replace Governor Patten.

Chinese and UK foreign ministers exchanged visits in the year following the CFA deal, but there were no substantial issues left to be discussed, let alone agreed. The reality was that China was taking charge in Hong Kong, and would do so on its own terms. Deng Xiaoping, who may not live to see 1 July 1997, would have it no other way.

Japan: Forward To The Past?

Japanese politics have come full circle. The unlikely coalition government, cobbled together in June 1994 from long-standing political rivals, defied repeated predictions of its imminent demise and survived throughout 1995. But Tomiichi Murayama, Japan's quiet, elderly Social Democratic Party of Japan (SDPJ) Prime Minister, rarely seemed to be enjoying his post and on 5 January 1996 he resigned, leaving the way open for the Liberal Democratic Party (LDP's) new leader, Ryutaro Hashimoto, to succeed him. Thus, two and a half years after losing power, the LDP has once again generated Japan's Prime Minister, albeit now as the head of a coalition government. The main opposition party, the New Frontier Party (NFP or *Shinshinto*), also elected a new leader, the arch-strategist Ichiro Ozawa. The competition between these two young, dynamic politicians may be setting Japan on course to a more lively domestic political scene and, ultimately, to a more active international role.

The air of impermanence and hesitancy that surrounded the Murayama government in 1995 did little to help it deal with domestic problems, such as rebuilding the earthquake-devastated Kobe region or resuscitating the broader economy which, battered by the high yen, continued to limp. A series of nerve gas attacks by, and continuing revelations about, the *Aum Shinrikyo* (Supreme Truth), an extreme religious cult apparently intent on setting up an alternative government, made ordinary Japanese concerned both about the state of their contemporary society and about the ability of their government to maintain law and order. The national mood turned inward-looking: perhaps the political and economic systems (including corporate lifestyles) that had seemed so essential to the success of post-war Japan were no longer as effective or relevant. But how exactly could these be changed? There were no easy answers.

Domestic preoccupations once again constrained foreign-policy initiatives in 1995, although the government demonstrated a discernibly firmer tone in dealing not just with the United States, but also with China. Japan's highest profile came with its role as host to the annual APEC Conference in Osaka on 18–19 November, but this display of regional leadership proved a mixed blessing. The Japanese would have preferred to keep a low profile, with painful memories of the past still clouding relationships in the present. All in all, this mixture of gloom and doom made 1995 a year which many Japanese would prefer to forget.

Surviving Against the Odds

The unholy alliance of the LDP and its long-standing socialist opponents, the SDPJ and a small LDP splinter party, *Sakigake*, held together in 1995 despite a series of crises. But Prime Minister Murayama only survived by compromising his party's most dearly held principles and accepting tacit LDP control. The SDPJ agonised over its policy changes and endlessly discussed revamping itself as 'a new party'. The overall effect was a steady loss of credibility in the eyes of the general public.

SDPJ politicians were not alone in losing public confidence. As low voting turnouts in the April 1995 local, and July Upper House (House of Councillors), elections showed, apathy towards politicians of all established parties was becoming pronounced. But symptomatic of a deeper malaise in Japanese society was the public questioning of its previous faith in the effectiveness of the bureaucracy and the instruments of order.

The devastating Kobe earthquake in January 1995 cast doubts on previous official claims of well-developed earthquake-proof construction techniques, and on the government's ability to respond effectively and quickly to such a crisis. A further blow to the national psyche occurred on 20 March, when sarin nerve gas was released into the Tokyo subway system, killing 11 commuters and incapacitating some 5,000 others. Within a few days, two further but smaller gas attacks occurred in Yokohama. The Japanese police suspected the *Aum Shinrikyo* of being behind the attacks, but had difficulty finding sufficient evidence to bring those responsible to court. At the end of March, the National Police Agency chief was assassinated outside his home. Basic Japanese assumptions about personal safety and security inside the country were thus badly shaken.

In mid-May, after much public criticism of their failure to bring those behind these attacks to justice, the police raided the *Aum Shinrikyo's* compound on the slopes of Mount Fuji and arrested the cult's half-blind leader, Shoko Asahara. During the subsequent months of investigation, details emerged that the more than 10,000 cult followers had been organised into a shadow government, with ministries, laboratories and stockpiles of weapons. The trials of Asahara and his closest associates, with no doubt new lurid revelations of their activities, are set to become a media highlight of 1996.

The cult was able to evade detection for so long in part because it was covered by the lenient post-Second World War laws protecting religious organisations adopted in reaction to the authoritarian control of the pre-war militarist government. In direct response to *Aum Shinrikyo's* attacks, the Diet passed new legislation in late 1995 tightening controls over religious groups and giving the authorities greater powers to investigate their finances. Although the shock of the attacks provided general support for this move, supporters of the former *Komeito* political party (now subsumed into the NFP), who still have strong connections with the *Soka Gakkai* Buddhist organisation, strongly opposed some of its provisions.

Public disappointment with the government's initially poor responses to both the Kobe earthquake and the nerve gas attacks (coupled with growing concerns about the lack of countermeasures to deal with the high yen) was reflected in the April 1995 local elections. Yet the opposition NFP made only modest gains. Instead, the headlines were stolen by two former television entertainers, Yukio Aoshima and 'Knock' Yokoyama, who ran as independents and were elected as governors of Tokyo and Osaka respectively.

The SDPJ had suffered the heaviest of the blows dealt to the coalition government parties in the local elections and the same pattern was repeated in the Upper House elections in July 1995. Under the existing electoral law, half of the Upper House's 252 seats were contested. The coalition's strength was cut back from 158 to 148, and the SDPJ lost almost two-thirds of the seats that it was defending. The NFP registered the most gains, winning 40 seats and emerging as the second-largest party in the Upper House after the LDP. But the clearest message from the elections was the record low turnout, the first time in post-war history that less than 50% of voters had bothered to go to the Upper House polls. Voters seemed unconvinced of any real differences in the parties' political platforms and they clearly distrusted politicians in general.

Murayama resisted calls for both his resignation and a general election, believing that maintaining the coalition would be the only way of giving the SDPJ sufficient time to regroup. The LDP, however, had decided that change was needed within their own leadership. Yohei Kono, the upright, but mild, leader brought in to restore the LDP's image after its 1993 debacle, was persuaded not to seek re-election as party president in September. Brushing aside a last-minute rival candidate, Ryutaro Hashimoto swept to victory. A flashy dresser and more forward and outspoken than Kono, Hashimoto had helped to ensure his rise to the top through his tough – but popular – stance in negotiations with the US over car parts earlier in the year. He took over the deputy premiership from Kono and bided his time before opting for the premiership itself.

The NFP in turn then felt the need to revitalise its leadership. Toshiki Kaifu, the former Prime Minister and belated convert to the anti-LDP camp in 1994, stepped down and Ichiro Ozawa easily defeated another former Prime Minister, Tsutomu Hata, in the run-off. Ozawa had been very much the strategist behind the scenes since he and Hata had worked together in 1993 to split the LDP. An ambitious politician who occasionally appears to his compatriots as almost too clever, Ozawa is also among the few Japanese politicians with any vision of the country's future. He has forcefully advocated political reform and faster deregulation at home, and a more active role abroad, in order for Japan to be considered a 'normal' country.

By early January 1996, Murayama had finally had enough and retired from the premiership, ensuring Hashimoto's promotion. Although the largest party in Japan, the LDP alone cannot command a parliamentary majority in either House, and needs coalition partners to govern. Hashimoto managed to keep the SDPJ and the *Sakigake* on board, offering the deputy premiership to the SDPJ's second-in-command, Wataru Kubo. But Hashimoto's forthright views on a number of key issues, notably defence and Japan's past, are certain to make the two coalition partners increasingly uncomfortable. Moreover, new Finance Minister Kubo, and by extension the SDPJ, has been left with the almost impossible political task of bailing out the banking *jusen* (housing-loan corporations) with taxpayers' money; Kubo has clearly been set up as the scapegoat for public anger.

The NFP's tactics are to amplify the differences between the LDP and its two coalition partners and then split them. The LDP, similarly, is trying to split the NFP, itself a recent amalgam of nine opposition parties and groups. In fact, neither side may have to try too hard, as the two 'coalitions' may anyway be heading for self-destruction.

The SDPJ did try to reconstitute itself in January 1996, but this amounted to little more than changing its Japanese name to coincide with the English name it had adopted several years earlier: it offered no significantly new policies. The SDPJ is still vulnerable to a split between those wedded to the old socialist ideals and those who are looking for a more reformist approach. The small *Sakigake* party is steadily moving closer to the SDPJ on policy issues and may well merge with its reformist faction in due course. Proselytising by *Sakigake* and unease with Hashimoto may well bring about the final split in the SDPJ, which in turn would bring down the LDP-led coalition. The NFP is hardly more stable. Barely a month after Ozawa's victory, the loser, Hata, established his own 'study group', traditionally the first step towards a new faction or even a new party. Thus the Japanese political game of 'changing partners' looks set to continue for some time.

Economic Blues

One of the major tasks facing Hashimoto was to restore Japan's economic health, but it was not immediately clear from his policy platform which aspects of the cautious and largely ineffective approach adopted by the coalition under Murayama he planned to change.

During the first half of 1995, the Japanese economy continued to stagnate. Even the Economic Planning Agency (EPA), which had repeated its unrealistically optimistic mantra that 'recovery is in sight' for months, acknowledged by the summer that the economy had stalled. Unemployment rose to a record 3%, industrial production declined and exports slowed. The Japanese, used to periods of a high yen rate (or *endaka*) in the mid-1980s, began to talk about 'super-*endaka*' as the yen rose from around 100 to the US dollar at the beginning of 1995 to a record high of 79 to the dollar in April. More and more companies found they had cut costs as far as possible. Some, such as the Nissan car company, took the drastic step of modifying the traditional employment-for-life system and began laying off workers. Others, led by the major electronic companies, accelerated their plans to relocate production to overseas sites, particularly in Asia.

Japanese confidence took a further battering during the summer as the medium-sized regional Hyogo Bank and a series of credit unions and small financial institutions declared bankruptcy. In addition, it was revealed that a rogue trader in Daiwa Bank had racked up losses of more than one billion US dollars. The credibility of Japanese financial institutions, and the Finance Ministry itself, were increasingly in question.

The Bank of Japan progressively cut interest rates to the barest minimum, while the Murayama government struggled to find ways of boosting the economy. In spring 1995, it announced a new five-year programme intended to cut regulations and red tape, but this had little short-term impact. In desperation, the government agreed in September to a huge reflationary package of ¥14.2 trillion, which gave a massive transfusion of money to public-works projects but provided no solution to the growing problem of bad debts in the financial sector. Its one attempt to deal with part of the financial sector's problems – a scheme to dissolve seven insolvent housing-loan companies and to cover part of their loans from public funds – proved highly controversial.

More by good luck than good judgement, the economy that Hashimoto inherited began, in early 1996, to show signs of life. The yen stabilised to around 105 to the dollar, giving manufacturers a breathing space, and the stock-market rallied to levels not seen for several years. The EPA was encouraged enough to claim that the economy was 'pulling out of the state of standstill'. False dawns have appeared before in the past four years of recession, but throughout it all Japanese industry has survived. It has certainly become leaner; 1996 will show whether it has become any fitter.

Reassessing Key Relationships

The Japanese had always expected 1995 to be a difficult year for diplomacy, given the inevitable fiftieth anniversaries of the end of the Second World War. Laying to rest the ghosts of the past would never have been easy, but Japanese politicians contrived to make it yet more difficult. With their leader in office, the SDPJ tried to seize the opportunity to ram an anti-war resolution through the Diet, but many LDP members balked at including a forthright apology for Japan's wartime actions. The result was an insipid Diet resolution in June which merely expressed 'deep remorse'.

The lengthy internal wrangling about the exact wording of the resolution only served to give the impression that Japan's apology was not genuine. Murayama, a lifelong pacifist, was embarrassed by the government's failure, and on the anniversary of Japan's surrender on 15 August he became the first Japanese Prime Minister to use the word 'apology' to the country's Asian neighbours.

This positive development was soon marred, however. Murayama himself caused controversy by supporting the legal validity of Japan's annexation of Korea in 1910, Cabinet Minister Takami Eto was forced to resign after extolling the benefit of Japanese colonialism for the subjugated nations – specifically Korea – and Ozawa accused China and South Korea of stirring up anti-Japanese sentiment for political reasons. Hashimoto's record of close collaboration with Japanese veterans' associations suggests that Japan's past will continue to haunt its present and future foreign policies.

The third item at the top of Hashimoto's agenda, other than the economy and the financial sector, was undoubtedly the state of Japan's alliance with the United States. In the first part of 1995 the focus had been on an increasingly abrasive economic relationship between them, with Hashimoto, then Minister of International Trade and Industry, taking the lead in standing up to US demands. A number of sectoral issues became contentious, but both sides concentrated their main energies on the car parts negotiations. The Americans went so far as to threaten to impose nearly $6bn in punitive tariffs on imported Japanese luxury cars. A last-minute compromise was struck on 29 June, and the Japanese managed to avoid a repetition of the 1991 agreement on semi-conductors that had delineated specific numerical targets for market-share for Japanese imports from the United States.

In the second half of 1995, however, the focus shifted to the security relationship, even though the Japanese and the US governments ironically found themselves united in wanting to justify and continue their security alliance. Events in Okinawa caused the most serious crisis the US–Japan alliance has faced since the massive protests against the security treaty's revision in 1960. In September 1995, three US servicemen were arrested and charged with raping a young teenage girl in Okinawa. The crime particularly incensed the Okinawans, who have long been resentful of the heavy US military presence on their islands. More than three-quarters of all US bases and nearly two-thirds of all US servicemen in Japan are located there. Large-scale demonstrations demanded the removal of all US bases from Okinawa. In support of this sentiment, the governor of Okinawa refused to sign documents requiring landowners to extend the bases' leases. He made it clear that he was not totally opposed to a US security commitment; he just wanted it moved to someone else's backyard.

The SDPJ had in the past led the fight against the US–Japan alliance, but Murayama now argued for its maintenance and both governments worked to curtail the damage. A minor revision to the Status-of-Forces agreement, which made it easier for Japanese authorities to arrest US servicemen suspected of crimes in Japan, was quickly agreed. Embarrassingly for the US, the Commander-in-Chief of the US Pacific Command, Admiral Richard Macke, was forced to take early retirement when he made an injudicious comment on the rape case. The Japanese persuaded the Americans to set up a review panel to discuss consolidating and reducing the US bases in Okinawa. The initial US position was that this did not imply an overall reduction in its forces stationed in Japan, only their redistribution away from Okinawa, but the weight of public sympathy throughout Japan for the Okinawan plight and the consequent anti-Americanism will make it difficult to negotiate significant relocation within Japan. The net result is certain to be a reduction in US deployments in Japan as a whole.

The difficulties the two governments faced were compounded by US President Bill Clinton's failure because of the US budget crisis to attend the November APEC summit meeting in Japan. Even before the Okinawan incident the two sides had been drafting a joint US–Japan 'security declaration' reaffirming their commitment to the security alliance. Although this declaration has only been postponed until Clinton's expected April 1996 visit to Japan, the delay has fuelled internal Japanese debate about the value of the alliance.

Since the end of the Cold War, Japanese public opinion has slowly cooled towards the security alliance; the Okinawan incident has now pushed support for its continuation into the minority. The problem facing both the Japanese and US governments is how to convince their publics that there is still a need for the alliance. The Japanese government tried to lead the way by adopting new defence guidelines in November 1995. Nearly 20 years after the first National Defense Program Outline, the second aimed to broaden the scope of Japan's security responsibilities to cope with the changed realities of the external world. The 1976 Outline's reference to Japan dealing with 'limited and small-scale aggression' was replaced with clearer procedures for coordinated US–Japanese operations and a commitment not just to defend the immediate Japanese islands, but also to take 'appropriate measures' to 'establish a more stable security environment' in the Asia Pacific region. The new Outline was therefore a realistic step forward and one which moved Japan closer in practice to the 'collective defence' concept which all Japanese governments claim is constitutionally forbidden. At the same time, the Self-Defense Forces (SDF) will be made more compact and more high-tech: a new rapid-reaction force will be established; the number of Ground SDF will be reduced with some divisions redeployed from Hokkaido to the more southerly islands of Japan; and more advanced weaponry will be purchased or developed.

The rationale of Japanese politicians and officials for these measures was the 'unpredictability and uncertainty' of the regional environment. Although not explicitly mentioned in the Outline, Japanese defence officials have made it clear that North Korea and China are the two major potential security threats in this context. The territorial dispute with Russia is no nearer solution, but the Japanese see Russian military capabilities in the Far Eastern region as in continuing decline. Japan is much more concerned about North Korea's newly developed missile capabilities and have doubts about its nuclear programme. Consequently, the defence budget proposed for 1996–97, which shows roughly a 2% increase compared with the less than 1% of the past few years, contains provision for a new anti-missile defence system, under the theatre missile defence programme with the US. While the Japanese twice gave food aid to North Korea, in all other aspects the Japan–North Korea relationship remained at a stand-off in 1995.

The greatest, and probably most significant, change in Japan's attitude is towards China. Not only did the Japanese take a tougher line on economic issues, such as refusing to renegotiate yen loan terms and pressing the Chinese to restrain textile exports to Japan, but they also expressed concern about the Chinese military build-up and Chinese actions in the South China Sea. China's insistence on continuing its nuclear testing brought a strong protest from Japan. Murayama was particularly disappointed that the tests continued, even after he had expressly asked the Chinese during his May 1995 visit not to undertake any more. As a result, Japan took the unprecedented step of cutting back its grant aid to China. Although the much larger yen loans were left untouched, the action had symbolic importance and would not have been possible without a distinct change in attitude.

Hashimoto, in his first parliamentary policy speech after becoming Prime Minister in January 1996, stressed that Japan should become more active in international affairs. His opposition rival, Ozawa, has also often talked about making Japan a 'normal' country within the international system. Unfortunately, despite the occasional signs of firmness and resolution shown over the past year, external contributions will still be at the mercy of domestic considerations for the immediate future, at least until fresh elections provide a new mandate.

The Slow Progress Of Multilateralism In Asia

The novel effort to construct a multilateral security regime in the Asia-Pacific was strengthened somewhat during 1995, but it still failed to deal directly with any of the region's major conflicts. There is still no regional constituency for a traditional balance of power – despite the potential for some of these conflicts to escalate – because, above all, the threat of Chinese hegemony would need to be addressed. It is not just a case of diverging strategic perspectives and a lack of political will; the Asia-Pacific nations have judged that turning to a balance of power is impractical in current circumstances. Thus the ASEAN Regional Forum remains the sole vehicle of common security. China, the United States and Japan, in particular, will have to use this embryonic multilateral security structure if they are to engage in any dialogue that might develop mutual restraint, perhaps through mutual economic interests.

The ARF is based on a security model pioneered by the Association of South-East Asian Nations. ASEAN's intention has not been to solve problems *per se*, but rather to create a regional milieu in which they either do not arise, or in which they can be readily avoided or managed. But whether ASEAN's distinctive security practice can be transposed onto a wider regional canvass, where conflicts may not necessarily lend themselves to the Association's quasi-familial approach, is in question. Also in question is whether what passes for dialogue amounts to more than diplomatic formulae.

ASEAN has sought to ensure its diplomatic centrality within the ARF, for which it took the formal initiative in 1993. Partly with a view to maintaining that centrality, the Association has begun to expand its membership. Some tension has developed, however, between its declared regional position and its interests, exemplified by the South-East Asia Nuclear-Weapon-Free Zone (SEANWFZ) Treaty, concluded at its fifth summit in Bangkok on 15 December 1995, and the wider interests of the United States and China which have raised differing objections to the Treaty's terms. In addition, Australia and Indonesia have departed from multilateralist practice by concluding a unique agreement on security outside both ASEAN and the ARF.

The ASEAN Regional Forum

The second working session of the ARF convened in Brunei on 1 August 1995. At this meeting, the inclusion of Cambodia, which had qualified for membership by acceding to ASEAN's Treaty of Amity and Cooperation, the Association's code of conduct, increased the number of members to 19. It was not the most propitious time for a meeting, for it took place against a background of crisis in Sino-US relations, of tension in Sino-

Japanese relations, and of strain in China's relations with ASEAN. ASEAN had not initially responded with any vigour when it became clear in February 1995 that the Chinese had introduced a naval presence on the Philippine-claimed Mischief Reef in the Spratly Islands. But in Hangzhou in April 1995, at a private dinner on the eve of a prearranged security dialogue between senior officials from the Chinese and ASEAN foreign ministries, the ASEAN officials set forth a strong and united view on the subject. This display of solidarity placed China in a position of relative diplomatic isolation as the ARF convened for its second working session. Moreover, at the end of July, Vietnam, which had earlier in the month re-established full diplomatic relations with Washington, was admitted as the seventh member of ASEAN.

The ARF meeting was guided by a novel Concept Paper which had been approved in advance. It set norms for the activities of the ARF, and provided an outline for its functional evolution in three stages: from initial confidence-building, through preventive diplomacy, to conflict-resolution mechanisms. At China's insistence, however, the notion of mechanisms was omitted from the Chairman's concluding statement. The Concept Paper stressed the need for a gradual evolutionary approach to managing regional security, and recommended two complementary approaches. The first drew on the practice and experience of ASEAN in reducing tension and fostering regional cooperation through informal processes. The second was to be based on concrete confidence-building measures. The Paper also stressed the pivotal and proprietorial role of ASEAN as 'undertaking the obligation to be the prime driving force of the ARF' while acknowledging the full and equal role of all participants. It then stipulated that the ARF's rules of procedure should be based 'on ASEAN norms and practices' and that decisions should be made after careful and extensive consultations by consensus, without voting. The Chair of the ASEAN Standing Committee would provide the secretarial support and coordinate ARF activities. Finally, the Paper recommended that the ARF should progress 'at a pace comfortable to all participants'.

The Concept Paper was an obvious effort to stamp ASEAN's proprietary mark on the ARF, and it naturally generated resistance from some participants. Significantly the Paper was not formally adopted in the Chairman's concluding statement. Instead, it espoused a number of proposals 'in the context of the Concept Paper'. In addition, although the statement reiterated that ASEAN would undertake 'the obligation to be the primary driving force', it pointedly preceded this with an assertion of 'full and equal participation and cooperation by all participants' without explicitly endorsing that exclusive undertaking.

Despite the measure of tension over ASEAN's so-called primary driving role, the meeting endorsed the substance of the Concept Paper's recommendations. It identified an essentially consultative security role which it defined in comprehensive terms beyond solely military aspects, while making clear that the ARF would not try to impose solutions. To that end, the language of the Concept Paper was adopted in stating that 'the ARF process shall move at a pace comfortable to all participants', while the collective approach to security was described as 'evolutionary, taking place in three broad stages, namely the promotion of confidence-building, development of preventive diplomacy and elaboration of approaches to conflicts' (as opposed to the development of a conflict-resolution mechanism). The ARF process was described as at the first stage only.

An *ad hoc* support group that would operate between sessions was charged with addressing confidence-building measures, while inter-sessional meetings were given responsibility for discussing peacekeeping operations and search-and-rescue coordination. Using different names for two evidently similar bodies was again intended to accommodate China's objections, in this case to any impression of continuous institutionalised activities. The only other concrete measure achieved at the ARF meeting was an agreement that member-countries would voluntarily submit an annual statement of their defence policies. It dealt with three pressing matters of regional security, calling for an immediate end to nuclear testing, urging North and South Korea to resume their suspended dialogue, and expressing concern about overlapping sovereignty claims in the region. The problem posed by Taiwan was not mentioned.

The ARF meeting was a useful opportunity for Chinese and US foreign ministers to try to defuse the tension between their two countries that had developed over Taiwan, and that attracted more press attention than the outcome of the ARF session itself. The meeting, and the Chinese presence, also served one of the founding purposes of the ARF which was to assure China's regular participation. Indeed, China has only begun to feel comfortable with the multilateral enterprise because ASEAN has been able to impose its own format and operating procedures. These procedures allowed China to apply a diplomatic brake to its own advantage. By way of concession to ASEAN in particular, expanded through the membership of Vietnam, China associated itself with the Chairman's statement on competing claims to sovereignty. That part of the statement also encouraged all claimants to reaffirm their commitment to the principles contained in relevant international laws and conventions as well as ASEAN's 1992 Declaration on the South China Sea which had called for all parties to settle their claims peacefully.

That China went out of its way to indicate its recognition of international law, including the 1994 UNCLOS, as a basis for negotiating a settlement of the Spratly Islands disputes was greeted as a welcome clarification of its legal position, even though Beijing had yet to ratify

international boundaries

MONGOLIA

RUSSIA

Sakhalin

Urup

Iturup

Kunashir

Shikotan

Habomai

Hokkaido

Vladivostok

NORTH
KOREA

*Sea
of
Japan*

Beijing

Pyongyang

Tianjin

Seoul

SOUTH
KOREA

Honshu

Tokyo

Kyushu

CHINA

Shanghai

*East
China
Sea*

Okinawa

Taipei

PACIFIC

TAIWAN

OCEAN

HONG
KONG

Hanoi

MYANMAR

L A O S

Vientiane

V I E T N A M

*Paracel
Islands*

P H I L I P P I N E S

Manila

Yangon

THAILAND

*South
China
Sea*

Bangkok

CAMBODIA

Phnom
Penh

*Camh Ranh
Bay*

Spratly Islands

BRUNEI

Strait of Malacca

M A L A Y S I A

SUMATRA

SINGAPORE

BORNEO

SULAWESI

0   500 km

0   300 miles

I N D O N E S I A

N

UNCLOS. China also indicated its willingness to discuss the Spratly issue with all ASEAN members, thus moving away from the rigid bilateralism on which it had insisted up to the end of the meeting.

The concession, however, was one of form only. China gave no indication of a willingness actually to address its insistence on sovereign jurisdiction which, if ever realised, would enable it to command the maritime heart of South-east Asia. Given that China was acutely concerned that the United States was engaged in a new practice of containment by seeking to renege on its recognition of a single China, Beijing may well have deemed it politic to soothe some of ASEAN's concerns. In so doing the Chinese allowed ASEAN's leaders to express satisfaction with the progress of the ARF. In addition, they were pleased with the tone struck in the relationships made between the 19 assembled foreign ministers. In the absence of anything more substantive, the ASEAN model appeared to have prevailed. But was it producing security?

Economic Cooperation

The third meeting of heads of government and economics ministers of the Asia-Pacific Economic Cooperation took place in Osaka in November 1995. Although its agenda had never explicitly included security matters, the common understanding that buoyant economies would strongly underpin regional stability provided an implicit link. In addition, the contact between political leaders was an added opportunity for private dialogue on bilateral tensions. The meeting, however, was marred in that respect because US President Bill Clinton was unable to attend. Moreover, US Secretary of Defense William Perry caused consternation and confusion over the regional division of labour for economic and security cooperation by reviving the notion of a security role for APEC. His proposal was duly repudiated by US Secretary of State, Warren Christopher, who reaffirmed the United States' understanding of the predominant security role of the ARF.

The most difficult problem the APEC meeting faced was that of implementing the Bogor Declaration of 1994. This had committed APEC members to free and open trade and investment no later than 2010 for industrialised economies and 2020 for developing economies. Well before the heads of government convened it had become evident that these goals were overly ambitious, given the vulnerability of a number of economies. To address this problem, an adroit formula was included in an Action Agenda which upheld the basic commitment of Bogor while permitting individual states to interpret it liberally. Article 8 of the Agenda stated: 'Considering the differing levels of economic development among the APEC countries and the diverse circumstances in each economy, flexibility will be available in dealing with issues from such circumstances in the liberalisation and facilitation process'. The need for such creative ambiguity did not dampen collective enthusiasm for sustaining the APEC process. Intra-regional trade and investment provided enough impetus to maintain the momentum of the undertaking; the next meeting will convene at Subic Bay in the Philippines in November 1996.

The Fifth ASEAN Summit

ASEAN began a process of 'widening' with the entry of Vietnam in July 1995, while Cambodia was granted observer status. In addition, Myanmar became the last South-east Asian state to accede to the Association's Treaty of Amity and Cooperation. The fifth meeting of ASEAN heads of government which convened in Bangkok in December 1995 was unique. Not only was there a prime minister from a communist state, Vietnam, in attendance, but the heads of government from Cambodia, Laos and Myanmar also sat in, making it the first occasion at which all ten South-east Asian states had been represented. A commitment was made to expand ASEAN to include all nations within the region by 2000. This commitment confirmed ASEAN's desire for 'widening', but raised the question of how far this could be matched by a corresponding 'deepening' based on a shared sense of regional values. The personal chemistry displayed at Bangkok, expressed in the willingness of the heads of government to take decisions among themselves – including one on annual informal summits – as well as by inviting India to become a formal dialogue partner, augured well for the widening process.

Also significant was the clear insistence on continuing ASEAN's leading role within its more limited regional security ambit, despite the institutional expansion of strategic horizons in the form of the ARF. ASEAN's emphasis on this point underscored the tension that has grown between its post-Cold War need to address regional security in a wider context than just South-east Asia, and the desire to uphold its separate corporate identity and exclusive position within the confines of the region. An expansion in membership would reinforce ASEAN's diplomatic influence and centrality within the ARF. The conclusion of the SEANWFZ Treaty in December 1995 was intended to manifest ASEAN's right to establish regional positions, particularly if endorsed by nuclear powers within the ARF. The Treaty had been under discussion for years, and had been pressed, in particular, by Indonesia which had taken its role as Chair of the Non-Aligned Movement very seriously. In a post-Cold War South-east Asia, the Treaty seemed incongruous, albeit with some symbolic significance.

The Treaty included the unexceptional undertakings not to manufacture, store or test nuclear weapons and not to allow any other state to use another signatory's territory to do so. It allowed individual signatories to grant access to the military aircraft and naval vessels of nuclear powers.

The Treaty also set out the geographic scope of its application as well as addressing the passage of foreign ships and aircraft. On these matters ASEAN failed to gain the endorsement of China and the United States whose signatures had been sought in a supporting protocol. China raised objections to the reference to the continental shelves and exclusive economic zones of signatories, which Beijing interpreted as prejudicing its claims in the South China Sea. The United States had reservations over possible impediments to the freedom of passage of its naval vessels through South-east Asian waters and to its aircraft through regional airspace. The US objections, in particular, were taken by the ASEAN nations as evidence of a sustained interest in deploying a regular military presence in the region. The failure to attract the endorsement of two key ARF members highlighted the tension between ASEAN's exclusivist desires and its wider approach to regional security, and the difficulty in reconciling them. This problematic issue raises the question of how long ASEAN will be able to control the agenda and direction of the ARF. Much will depend on whether the major Asia-Pacific powers see continued utility in the Forum. For its part, China appears content with the ASEAN model because it is predicated on consensus and because the Association's diplomatic centrality is much preferred to that of either the United States or Japan.

Australian–Indonesian Bilateralism

The surprise announcement in December 1995 of an unprecedented security agreement between Australia and Indonesia was in sharp contrast to the great multilateral occasions in Brunei, Japan and Thailand. It came on the eve of the ASEAN summit and was apparently reached without any prior consultation between Indonesia and its regional partners. More unusually, it appeared to come as a great surprise to both Indonesia's and Australia's foreign ministries. Although Australia had long supported multilateral approaches to security, the agreement of a treaty status made very good geopolitical sense in Canberra, even though it affronted liberal opponents of Indonesia's brutal occupation of East Timor. Australia's irritation at ASEAN's claim to a primary driving role within the ARF may also have played a part in its decision.

That Indonesia joined in this agreement was even more astonishing. Since independence in August 1945, its governments have practised a form of non-alignment described as an independent and active foreign policy which has repudiated alliance formation. Moreover, it had presented its chairmanship of the Non-Aligned Movement in 1992–95 as a great diplomatic triumph. Yet, although one of the founders of ASEAN, Indonesia had from time to time been frustrated with its limited scope and constraining consensus. It also felt uncomfortable within the ARF because of the evident greater importance of the United States, China and Japan.

The security agreement between Australia and Indonesia owed much to the personal relationship between then Prime Minister Paul Keating and President Suharto, encouraged by Suharto's growing concern about China's rising power. The agreement has the spirit of an alliance. It states, *inter alia*, that the two parties would consult 'in the case of adverse challenges to either party or to their common interests, and, if appropriate, consider measures which might be taken either individually or jointly and in accordance with the processes of each party'. Although Indonesia took great pains to assert that the basic principles of its foreign policy had not been compromised, there can be no doubt that the agreement, if not a full-blown alliance, represents a turning-point in the foreign relations of the Republic which has never before entered into a formal security arrangement.

Is There Utility in the Asian Way?

The question raised by the Australian–Indonesian agreement is whether it should be seen as a complementary bilateral arrangement or whether it signals a lack of confidence in the embryonic structure of multilateralism in Asia-Pacific. The multilateralist approach has been enhanced in institutional form by following in the main the model and practice of ASEAN. Indeed, ASEAN has been determined to locate the political centre of gravity of Asia-Pacific multilateralism within South-east Asia. But, the more critical issues of regional conflict are to be found in North-east Asia. One such area of conflict, Taiwan, is excluded from the ARF by its terms of reference. Moreover, the problem of the Korean peninsula is being addressed governmentally on a bilateral basis, while there is no North Korean representation as yet in the ARF. There is only an informal multilateral dialogue for North-east Asia which neither excites the relevant governments nor has any representation from North Korea.

In the case of the South China Sea, which spans and joins South-east and North-east Asia, the ARF has had to be content with diplomatic gestures from an irredentist China, hoping that by becoming comfortable with its ASEAN-like process China will help to build the foundations of a new regional security architecture. But if all that the ARF can expect, or is willing to live with, are gestures of the kind that China has made over the South China Sea, the question arises: is engagement on ASEAN's terms enough? The standard answer in the Asia-Pacific is provided by another question: what is the alternative where conventional multilateral balance-of-power arrangements are not readily available, in part because they are deemed more likely to provoke than to protect? This standard answer may not be sufficient, however, if China is increasingly seen as a threat. Asian nations may then be willing, as Australia and Indonesia appeared to be, to strike out on other paths, including the return to a balance of power, even if the balancer must be the United States.

Unease In South Asia

Most South Asian nations, including the dominant power, India, were increasingly obsessed with domestic politics in 1995. India and Pakistan managed to survive another year of uneasy peace, although there were signs that attitudes were hardening on both sides. An entire year passed without any top-level contact between the two neighbours. In fact, except for some unavoidable meetings at multilateral forums, there was almost no official political contact between the two countries at any level.

India, Pakistan, Bangladesh, Sri Lanka and Nepal were all convulsed by domestic political upheavals in 1995. The major concern of their leaders, most notably in India, was therefore self-preservation. India's Congress Party government, led by P. V. Narasimha Rao, continued to battle against heavy odds. In Pakistan, ethnic violence in Karachi, the nation's main port and financial capital, dealt a serious blow to the economy and threatened greater political instability.

In Bangladesh, the opposition boycotted the political process, refusing to take part in a general election ordered by President Khaleda Zia. When it turned down her offer to hold another election in place of the one she had won in February 1996, which had generated widespread allegations of vote-rigging, the political stalemate acquired dangerous dimensions. Sri Lanka's President Chandrika Kumaratunga won a spectacular military victory in December 1995 against the Liberation Tigers of Tamil Eelam (LTTE) guerrillas when government forces captured the rebel stronghold of Jaffna in the north. But she was unable to translate this into political gains capable of resolving the ethnic conflict. As yet another elected government fell in Nepal the palace intervened, but in an entirely constitutional manner, preparing the way for another election in the not-too-distant future.

With so much happening internally, it was no surprise that regional relations, and foreign policy in general, were put on the back-burner. The only significant exception was the nuclear issue. This became a major concern in India as the Nuclear Non-Proliferation Treaty (NPT) review conference in April and May 1995 struggled with the question of extending the Treaty indefinitely. When that was finally agreed, there was increasing pressure, particularly from the US, to secure an early agreement on the related Comprehensive Test Ban Treaty (CTBT). Leaders of all political parties, with elections in their sight, adopted tough positions on the issue, linking it with national prestige, honour and sovereignty.

India's Shaky March to General Elections

India's tenth general elections, scheduled for 27 April and 4, 11 and 18 May 1996, will probably be the most complicated in the country's post-independence history. They are expected to be the first post-1947 elections in which no member of the Gandhi–Nehru family will stand, or even play a major role. In the past, the 'first family' of Indian politics automatically created an election-eve polarisation with extreme views expressed across ideological lines. Without the Gandhis this polarisation has disappeared. Instead, three clear groups are likely to emerge at the polls: the Congress Party and its allies, mainly smaller regional parties; the Hindu revivalist *Bharatiya Janata* Party (BJP), along with its own regional allies; and the increasingly powerful National Front–Left Front alliance. None of these groups are likely to win anywhere near the number of seats in the *Lok Sabha* (House of the People), the lower house of India's bicameral legislature, required to form a government. The only reasonably certain prediction is that India is in for a period of coalition rule.

But another set of developments, even more important than this three-way split, will have a long-term impact on India's polity and prospects for stability. In 1995–96, an unprecedented level of corruption was exposed in Indian politics, bringing in its wake an equally new trend of judicial activism. The *hawala* scandal, as it is popularly known, has caused massive damage to the credibility of India's political system and its politicians. At least eight serving union ministers and an equal number of prominent opposition leaders have been aggressively prosecuted on charges of bribery.

In the subcontinent, *hawala* is loosely synonymous with money-laundering. The region has a long history of exchange controls, and its elites, particularly its politicians, have traditionally kept their illegal incomes abroad. As a result, a well-oiled network of foreign-exchange racketeers has emerged. In effect, by paying a middleman any amount in local currency – and by paying a premium – the equivalent in hard currency can be accessed abroad. If bribes in hard currency are accepted abroad, the *hawala* operators pay back the equivalent in local currency at home. This latest scandal involves at least 115 senior government officials and politicians, alleged to have used S. K. Jain, a well-known fixer in New Delhi's seamy, subterranean network of businessmen and politicians, to process their bribes.

The roots of the scandal were uncovered in a chance raid on a Kashmiri militant hideout in the walled city of Delhi in summer 1991. The captured militants led the police to Jain, who was suspected of transferring foreign funds to the militants through *hawala* channels. But when searching a farmhouse on the outskirts of New Delhi, the police also found a diary listing sizeable sums of money against familiar-sounding initials. This list was soon discovered to be a 'who's who' of Indian politics, and those involved were suspected of being on Jain's payroll. Left to themselves, the police would have buried the case there and then. But word got around and a group of lawyers, a journalist and a newspaper cartoonist filed a public-interest suit, seeking the Supreme Court's intervention in the case.

An unprecedented frenzy of judicial activism ensued, with charge sheets filed against nearly two dozen of those listed in Jain's diaries. They included several of Prime Minister Rao's Cabinet, the President of the BJP, Lal Krishna Advani, and the President of the break-away Congress Party, Arjun Singh, who had claimed the support of Rajiv Gandhi's widow, Sonia. Rao himself was embarrassed because, even though his name did not figure in the diary, Jain told investigators that he had visited Rao's house within a week of Rajiv Gandhi's death and handed him large sums in cash.

This election-eve scandal will have far-reaching implications for India's political fortunes, alliances and even the entire political system. Until the scandal broke, the BJP was widely seen as the front-runner in the electoral race. It was believed that while the Party would lack a simple majority, it would emerge as the single largest group on its anti-corruption and ultra-nationalist platform. But with Advani facing charges of collecting bribes from someone who also financed Kashmiri Muslims, the effect of this platform was destroyed.

The scandal could have an even deeper impact on the Indian people's faith in the system of electoral democracy. The scandal was forced into the open mainly because the Supreme Court directed the Central Bureau of Investigation (the Indian equivalent of the FBI), which was investigating the case, to report directly to it instead of to Prime Minister Rao. Under the Indian Constitution, this is the function of the executive and not the judiciary.

At a different juncture the politicians would have tried to turn this improper action to their own advantage. The difficulties they could expect, however, were demonstrated when a judge at one of Delhi's district courts, while turning down bail for one of Rao's Cabinet ministers charged with harbouring terrorists in his house, called parliament a 'fish market' and a 'den of thieves'. Parliamentarians exploded in righteous anger. But several opinion polls conducted at the same time must have given them pause for thought. They showed overwhelming support for judicial activism, while more than 80% of the respondents expressed contempt for politicians.

Since the forthcoming elections are expected to produce a three-way split, shifting coalitions and, therefore, instability, the public's mood of growing disenchantment with the political system has raised considerable concern. There was intense widespread debate on whether the country needed, and deserved, a better political system. The army, apolitical as ever, kept its distance. But a continuing sense of instability, along with the

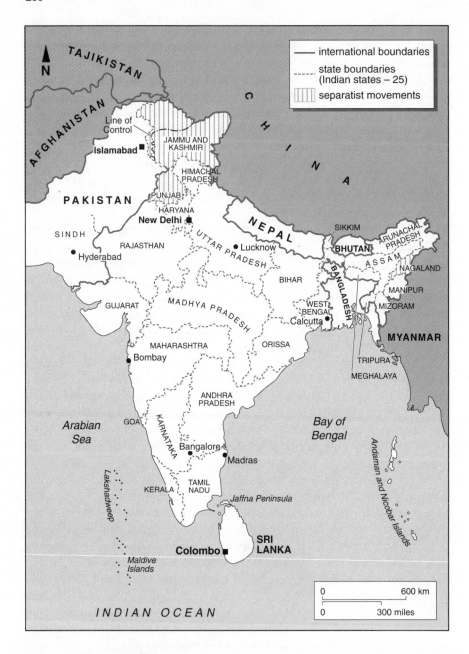

diminishing credibility of democratic institutions, threatened to weaken the nation and could ultimately encourage divisive tendencies in trouble-prone states and regions.

With the top leaders of almost all major national political parties, except the Left Front, involved in the *hawala* scandal, none of these parties could make corruption an election issue. There were still differences over economic policy, despite broad support for deregulation. But with annual inflation at a decade low of nearly 4.5%, a healthy balance-of-payments and foreign-exchange reserves position, export growing at more than 25% and industrial production at 11%, it was also difficult to turn the economy into the main election issue. Uncharacteristically for Indian politics, the focus shifted to national security.

National Security and the Nuclear Issue

In 1984, after his mother, Indira Ghandi, was assassinated and after the Sikh revolt, Rajiv Ghandi succeeded in converting widespread national insecurities into a massive mandate. Since then, however, no party has been able to make national security or foreign policy an election issue since all parties of any consequence have basically agreed on both foreign and security policies. In this sense, 1996 could mark a watershed with clear differences emerging, particularly between the BJP and the other parties.

Three key factors have brought about this shift. First, the BJP blamed all the continued troubles in Kashmir on the ruling party's 'soft' policies. Second, the BJP was able to exploit a curious incident in December 1995 when a Russian-made An-26 transport aircraft of Ukrainian registration, flown by Latvian pilots, was forced to land at Bombay by Indian Air Force MiGs. It was found to be smuggling several hundred assault rifles, rocket launchers and other armaments into eastern India. Government investigators claimed that the arms were meant for a controversial and mysterious fringe sect, the *Ananda Marg* (Path to Joy). In the past, the bearded, ochre-clad *Ananda Margis*, mostly active in the eastern, Marxist province of West Bengal, had acquired notoriety through acts of violent protest, attempts at self-immolation and strange religious practices including worshipping human skulls. The Rao government was unable to explain convincingly why the sect was attempting to import, through international syndicates, enough automatic weapons to arm an infantry battalion. The incident dominated headlines for several days and figured prominently in parliament, allowing the BJP to present it as another example of the government's 'failures' on the security front.

Third, and most significant of all, the BJP made capital out of the indefinite extension of the NPT, the contentious debate over the CTBT, and firm signs that the US was planning to renew its arms links with Pakistan. The opposition party took an aggressive stance on each of these issues and as US pressures to fall into line over the CTBT increased and public opinion in India hardened, the BJP stood to gain. An opinion poll conducted by *India Today* showed that the vast majority of Indians supported aggressive nuclearisation.

It had been widely accepted in India that during their meeting in 1994, President Rao and US President Bill Clinton had reached an amicable understanding. The US would not press India too hard on the sensitive issue of Kashmir and its forces' human-rights record there, while India would refrain from any aggressive deployment of *Prithvi*, its short-range missile, avoid further tests of the intermediate-range ballistic missile *Agni*, and generally not cause trouble at the NPT extension conference. Unfortunately, the indefinite extension of the NPT provoked the vociferous lobby of India's nuclear hawks. The government could probably have ridden out this storm, but it found itself cornered when the Clinton administration's effort to seek a one-time waiver on the Pressler Amendment, which banned US aid to Pakistan if it crossed the nuclear threshold, was granted by Congress.

In South Asia, a relationship with the US is still seen as a zero-sum game. Thus Indian public opinion was quick to translate what it interpreted as yet another concrete example of a US 'tilt' towards Pakistan into a deliberate affront to India. The timing of this development – just as discussions at the October 1995 Geneva-based Conference on Disarmament (CD) were entering a significant phase – was particularly inauspicious. Indian public opinion was further hardened by the aggressive manner in which US officials pushed for an early signing of the CTBT.

In winter 1995, the *Washington Post* quoted claims by intelligence sources that India was readying its 1974 test site in the Rajasthan desert for an imminent nuclear test. New Delhi denied the reports, but in an environment where the US is widely viewed with suspicion and nuclear weapons are seen as a vital indicator of national prestige, these denials evoked scorn. India's External Affairs Minister, Pranab Mukherjee, tried to satisfy both constituencies by claiming that India was not planning to hold a test, although it was fully capable of doing so.

This ambiguity did not pass muster at home. The nuclear debate at Indian institutions and in the media in 1995 had acquired an unusually sharp edge and the doves were being increasingly sidelined. Alarmed, the government tried to assuage public opinion by holding another test-firing of the *Prithvi* in January 1996 and media reports claimed that the missile had actually been deployed with front-line artillery regiments. The latest *Prithvi* version was also paraded on Republic Day, 26 January 1996. Yet it was evident that the present Indian government would find it impossible to adopt a particularly conciliatory posture on the CTBT, at least in the short run, while the BJP was exploiting the issue as an election gambit.

With both the US and India bracing themselves for elections, the differences over the CTBT are threatening to become a major irritant in their relations. India now links the CTBT with global disarmament. It argues that while the Treaty is acceptable on its own merits, in its present form it merely perpetuates the nuclear caste system. Until the nuclear powers renounce all their nuclear weapons, or agree to a firm timetable for doing so, India will not adhere to a CTBT. In the US, however, this is seen as sophistry. Many US officials portray the new Indian stance as a complete *volte face* from its earlier co-sponsorship of a complete test ban at the UN, beginning with Jawaharlal Nehru in the 1950s. New Delhi, however, argued that while it still supported a complete test ban in principle, the indefinite extension of the NPT had changed the situation fundamentally.

The issue is certain to sour relations between India and the US for some time to come. While it does not perhaps have the same emotional associations in the US as in India, it nevertheless still has resonance in US domestic politics. In India, the hawkish nuclear lobby has virtually hijacked the debate, although arguments in favour of defying the world were far from convincing. The dominant sentiment underlying this mood is to defy the US, which is seen yet again as a friend of Pakistan and hostile to Indian interests. India's government tried to pander to this sentiment by publicly repeating its assertion that its nuclear programme has not been abandoned. The anti-US mood was also evident in the manner in which S. B. Chavan, India's ageing and conservative Home Minister, lambasted the US, accusing it of having ulterior motives towards Kashmir.

By adopting such an activist posture on the nuclear issue, India solved some problems for Pakistan, which could wait for its neighbour to make a move. The US media also carried intelligence leaks that Pakistan was readying its Chagai hills test site in north-western Baluchistan and was likely to respond within days to an Indian test. But a more discordant note was sounded by reports that Pakistan was once again involved in the clandestine purchase of nuclear material from China, a move that led to only mild US sanctions against Beijing.

Other Defence Issues

India made no major defence purchases during 1995, but its forces appeared to be pressing for more. The defence budget was raised marginally ahead of inflation. The armed forces also evaluated options on self-propelled artillery, a multi-role aircraft and better electronic countermeasures. Only an escalation of tensions with Pakistan, however, would move the government to spare money for new purchases, including the long-delayed acquisition of an advanced jet trainer, in the near future.

Part of the squeeze on defence funds came from the report of India's Fifth Pay Commission which was examining the wages of all government employees, including defence personnel. It seemed clear that even an increase of 10% in the defence budget, after adjusting for inflation, would be consumed by these wage rises. Although India was having difficulty finding money for new purchases, Pakistan moved closer to finalising a deal with France for nearly 40 *Mirage*-2000s and other armaments, including *Exocet* missiles. This led to strong Indian protests and the visit of a high-level Indian delegation to Paris was cancelled. But India was more concerned about the Brown Amendment resuming the US arms supplies to Pakistan, even if it was to be a one-time waiver. India's concerns were particularly focused on the *Harpoon* missile/P-3C *Orion* combination that this package contained.

Escalation in Kashmir

Both the war of words that India and Pakistan have engaged in over Jammu and Kashmir and the level of violent exchanges along the border escalated in 1995. Although the two armies frequently exchanged fire in the disputed territory, military casualties were kept to a minimum. In the worst case, 20 civilians were killed in January 1996 when rockets struck a mosque on the Pakistani-controlled side. India denied Pakistani allegations that it had fired the rockets, alleging in turn that fire from a Pakistani battery had apparently strayed off target. Some Pakistani military leaders talked of revenge, but the situation was finally defused at the local level.

Within Indian Kashmir, however, unabated trouble reached a dangerous point in summer 1995 with the destruction of an important Sufi shrine, *Charar-i-Sharif*, in May. When a band of militants, led by a mercurial commander called 'Major' Mast Gul, occupied the shrine, the Indian Army besieged it. In the ensuing attack the shrine, of exquisitely carved walnut, was destroyed by fire. Some of the militants were killed, but the destruction of the shrine and Mast Gul's escape caused international embarrassment. The incident was particularly damaging for New Delhi, coming at a time when even impartial observers and diplomats visiting the Kashmir valley had begun to feel an increase in popular disenchantment with the never-ending militancy. India derived some minor consolation later when Mast Gul reappeared in Peshawar, Rawalpindi and Karachi, leading victory processions of followers armed with AK-47s, and thus embarrassing Pakistan.

The destruction of the shrine, followed by the kidnapping of four Western tourists by the fringe fundamentalist group *Al Faran*, set back efforts to restore the political process in Kashmir. Despite the beheading of a Norwegian hostage, New Delhi stubbornly resisted the group's demand for the release of several prominent jailed militants in exchange for the other hostages.

By far the greatest blow came when the Rao government's decision to hold elections in the troubled state was upset by Central Election Commissioners who, after visiting Kashmir, declared that it was not possible to hold a free and fair poll. Rao's strategy had long been to hold shotgun elections to ensure a government of local people, however unrepresentative. But this failed before he could initiate it.

Another setback to the government was its failure to persuade the only moderate and pro-Indian Kashmiri party of any consequence, the National Conference led by Farooq Abdullah, to participate in the political process. Abdullah, the son of the late Kashmiri strong-man Sheikh Abdullah (who was party to the state's accession to India), set out preconditions on constitutional autonomy that were impossible for the Rao government to accept in a general election year.

There were, however, some positive developments in India. The level of combat was lower, trade and commerce were less paralysed than in previous years, and some moderate leaders of the all-party *Hurriyat* (Freedom Conference) expressed their willingness to talk to New Delhi. India is likely to persist with its policy of waging a low-key war against militants while trying to hold elections. Should the BJP win the forthcoming elections in India, however, this could quickly change as it is likely to adopt a much tougher posture.

Entering the Market

Despite the usual election-year uncertainties, some trends became clearer. It was evident that deregulation and economic reform had come to be accepted, though in different forms, by all parties. Two high-profile events stand out. First, the right-wing BJP–*Shiv Sena* (Army of Shiva) coalition that came to power in Maharashtra, India's most industrialised state, cancelled a billion-dollar power project being set up by the US multinational company Enron, alleging that the previous Congress government had approved it in return for favours. It raised other objections, ranging from environmental destruction to bankrupting the state through an outflow of foreign exchange. Yet, despite all the noise, the project was approved by the new Maharashtra government. During the negotiations leading up to the new approval, Enron made substantial concessions. Nevertheless, that the project was re-established so quickly was significant.

Second, India took its first steps towards privatising its infrastructure sector by allowing parallel, private-run telecom networks. Cellular phones quickly appeared in the metropolitan towns, but the opposition delayed the privatisation of basic telecom services by blocking the entire winter session of parliament and then filing a petition in the Supreme Court. In February 1996, moving with remarkable alacrity, the Court disallowed the anti-privatisation writ. These developments indicated a changing attitude in the country, and supported the belief that whatever government is elected in spring 1996 will continue a policy of economic reform.

Pakistan: Benazir Strides Ahead

Prime Minister Benazir Bhutto managed to consolidate her position in Pakistan's political life in 1995 with the help of two successes, one in the foreign and one in the domestic sphere. The first developed from her long-awaited visit to Washington in April 1995. Even if the carefully orchestrated domestic media attempted to make too much of it, the trip was a great success. Bhutto managed to convince President Clinton that the Pressler Amendment was unfair to Pakistan. In particular, she argued, it was unfair of the US to continue to refuse to deliver military hardware, including F-16 aircraft for which Pakistan had already paid.

This started the process that ultimately resulted in the Brown Amendment being passed – a significant accomplishment on its own. But more fundamentally, the visit had broken the deadlock in relations between the old allies. To some extent, the visit reminded Americans of their old security relationship with Pakistan, and Washington's rediscovery of its old Islamic friend in a strategic part of the world was Bhutto's most important achievement.

In the context of Pakistani politics this fresh start with Washington has implications far beyond foreign policy. In Pakistan, the army has the final say in the power troika of the army chief, the president and the prime minister. Analysts there generally believe that the army allowed Bhutto to return to power because it desperately needed new arms supplies from the US and believed that only Bhutto could break the ice with Washington.

Pakistani–US relations had been frozen during the rule of Bhutto's predecessor Nawaz Sharif, with Washington even coming close to declaring Pakistan a state supporting terrorism. Bhutto, on the other hand, is considered in Washington to be a modern, pragmatic and moderate leader. Her successful visit further helped convince the army of her utility and, in turn, led to her consolidation in power despite the troubles in Karachi and persistent opposition campaigns.

The second crucial development was the smooth installation of a new army chief. Given the crucial role of the army in Pakistan's power structure, the retirement of an army chief and the selection of his replacement is always of great interest and frequently a matter of fierce in-fighting. Traditionally, chiefs have been reluctant to quit. But not only did General Abdul Waheed Kakar retire gracefully in December 1995, he has thus far stayed out of politics. His successor, General Jehangir Karamat, has evoked no controversy and he too has a reputation for being a tough, no-nonsense professional with no interest in domestic politics. The importance of this for a civilian prime minister cannot be over-estimated.

In addition to increased violence in Karachi, another threat to overall stability emerged in the form of growing, violent fundamentalism. In November 1995, a car bomb wrecked the Egyptian Embassy in Islamabad, killing several people. This followed the killing in Karachi of two US Consulate employees. In both cases, right-wing Islamic fundamentalists are believed to have been involved.

During the Afghan-Soviet war that began in 1979, the loosely administered tribal territory around Peshawar had in effect developed into a *jihad* university. Scores of clerics ran their own centres and training camps producing holy warriors. With the end of the *jihad* in Afghanistan in late 1988, they looked for other 'causes'. Army officers who planned to overthrow the democratic, pro-US and therefore 'anti-Islamic' government, had expected support from this constituency. It appears for the moment that Bhutto has contained the threat. But much depends on the goodwill of the army which, in turn, depends on how popular she continues to be in Washington. She could be out of office if US arms shipments (from the old, pre-Pressler purchases) do not begin to land at Karachi in 1996.

Sri Lanka: Military Victory, Political Stalemate

President Chandrika Kumaratunga's peace offensive in late 1994 was barely three months old when it became apparent that Sri Lanka was heading for another round of ethnic fighting. But the level and intensity with which the new President's army renewed its offensive against the separatist LTTE surprised most regional analysts. So, too, did the level of success that the army achieved. In an uncoordinated offensive that defied the guerrilla ambushes, land-mines and even the north-eastern monsoon in heavily wooded terrain, the government managed to wrest control of all the major urban centres in the Tamil-dominated northern and eastern regions of the island. These included the city of Jaffna, the capital of the northern province and the LTTE's headquarters.

This was a massive blow to the rebels and was achieved with relatively little cost on the ground. This time the LTTE did not defend its headquarters as it had when the Indian Army tried to seize it in 1987. The guerrillas tried, instead, to balance the score by attacking government forces in the air and on the sea, thereby adding an entirely new dimension to the ongoing internal conflict.

Instead of trying to halt advancing Sri Lankan Army columns, the guerrillas shot down three military transport aircraft (An-26/32), exacting a heavy toll. Another transport aircraft, an HS-748, crashed near Colombo shortly after take-off and was believed to have been downed by the LTTE as well. What weapons the Tigers used has not been conclusively established, although the Indian media were quick to conclude that they were *Stinger* missiles of Afghan war vintage, bought from Pakistani arms smugglers. At the same time, the LTTE mined and destroyed three Sri Lankan naval vessels. These attacks were carried out by Black Tigers, the LTTE's dedicated suicide squad.

If losses had been confined to these aircraft and ships, the Sri Lankan government would perhaps have absorbed them all with a certain complacency, given the massive territorial gains it had made. But the Tigers managed to strike in the heart of Colombo far too often for the government just to sit back and savour victory. The large oil storage facility near Katanayake International Airport at Colombo was set ablaze. Two more major suicide bomb attacks, including one that destroyed the Bank of Ceylon in January 1996, underlined the fact that while Jaffna may have been won, the LTTE was far from defeated.

The attacks in Colombo heightened insecurity within Sri Lanka, and also among prospective foreign investors and tourists. As a result, Australia and the West Indies refused to play their World Cricket Cup matches in Sri Lanka scheduled for 17 and 25 February respectively. The matches had been sponsored jointly by India, Pakistan and Sri Lanka, probably the only time the South Asian neighbours had managed to collaborate in a long time.

While she was hailed for her military victory and emerged much stronger in domestic politics as a consequence, Kumaratunga also realised that relying purely on a military solution to the ethnic problem had its limitations. The LTTE had withdrawn to the jungles. Since it had carefully avoided taking on the Sri Lankan Army frontally, much of its strength and top leadership, unlike in 1987, were intact. It could safely be predicted that the LTTE would continue to carry out attacks in Colombo, Sinhala population centres, neighbouring Tamil areas, and on other economic targets. The ethnic divide is destined to continue to blight prospects for stability in the Island republic.

President Kumaratunga has not yet indicated what her future strategy will be. But she has already established herself as an activist leader. It is reasonable to expect her to follow her limited military victory with yet another peace offering which will probably include a considerable devolution of powers. Her earlier offer had been a reform package which proposed transforming Sri Lanka into a 'Union of Regions', with a major enhancement of the powers available to the Regional Councils that exist in each of the country's provinces. Kumaratunga seems to believe that the loss of Jaffna is such a blow to the LTTE's prestige among the Tamils that the Tigers will now have difficulty finding new recruits, as well as sources of funding, and thus will ultimately sue for peace. But few who have studied the LTTE share that optimism.

Regional Equations: Static and Stubborn

Hopes that regional relations could be improved through multilateral institutional arrangements were dampened yet again in 1995. Unlike the willingness of the South-east Asian states to make effective use of their regional body, the Association of South-East Asian Nations, the South Asian nations remained stubbornly resistant to change. Efforts to generate some life in their regional grouping, the South Asian Association of Regional Cooperation (SAARC), proved futile yet again. India hosted what was intended to be the SAARC annual summit in May 1995, but Pakistan's decision to send only its President, who is more or less a constitutional figurehead, reduced the meeting to one that could make no real advances.

Some progress was, however, made in the area of trade when a South Asian Preferential Trade Agreement moved closer to becoming a reality. Yet here, too, the bilateral India–Pakistan differences cast its shadow. India had accorded most favoured nation status to its smaller neighbour, but Pakistan refused to reciprocate; this brought loud protests from India, if not any retaliatory action as yet.

As has now become the norm in South Asia, there was no resolution in 1995 of the question of what is bilateral and what is multilateral or regional. Pakistan, for example, insisted on making bilateral issues part of the SAARC agenda. India, as always, opposed this with equal vehemence, and the so-called SAARC summit was yet again reduced to a farcical diplomatic jamboree. Until India and Pakistan can solve their bilateral problems, SAARC is destined to remain just that.

Africa

After the dramatic events of 1994 – the first post-apartheid South African elections and the genocide in Rwanda – Africa seemed quiet in 1995. But any suggestion that its absence from Western television screens and newspaper front pages meant that all was well was an illusion. 1995 was a crucial year, and 1996, with some 18 African states promising to hold elections, will be even more important. These elections are the first democratic tests of the governments that emerged from the wave of multiparty democracy that swept the continent in 1990–91. Just how deep has this democratic spirit penetrated?

In West Africa, four military governments are in power. Three of them (Nigeria, Niger and Gambia) have drawn up timetables for their transition to civilian government, but the scheduled ends are too far in the future to satisfy aid donors. Coup attempts elsewhere on the continent demonstrate that the people's desire for democracy has not dispelled the military threat. In other African countries, political leaders have learned to manipulate elections and, in several cases, opposition parties have boycotted them. Indeed, in Africa a concept of 'donor democracy' is emerging – just enough democracy to keep donors happy.

Very few African politicians have willingly accepted the spirit of democracy. In East Africa, for example, two political systems have been developed locally: Uganda's no-party democracy, and Ethiopia's ethnic democracy. Both involve ballots, but not conventional Western concepts of freedom of association. Meanwhile in Angola, the site of Africa's longest-running war, nothing has changed. The United Nations demobilisation stations are almost empty and the process is months behind schedule. It is clear that Jonas Savimbi, the *União Nacional para a Independência Total de Angola* (UNITA) leader, has no intention of trading his army for a seat in government. He will instead hold out for both, using, as ever, his dual-track policy of diplomacy and war.

On the economic front, Africa is changing, but there is no consensus on whether these changes will benefit the people or simply profit foreign investors and a narrow domestic elite. Those countries that adopted International Monetary Fund (IMF) and World Bank-sponsored reforms had growth rates slightly, but perhaps not significantly, higher than those that did not. Whether these reforms are 'working' or not, ordinary people in Africa continue to face a long tough struggle if they, or their children, are ever to emerge from subsistence levels of existence. Meanwhile, the ability of governments to provide adequate health and education diminished as state expenditure was cut; and Western aid levels are expected to drop sharply in the next few years. Military expenditure, however, continued to rise and arms, both for government and rebel movements, still flowed into the continent.

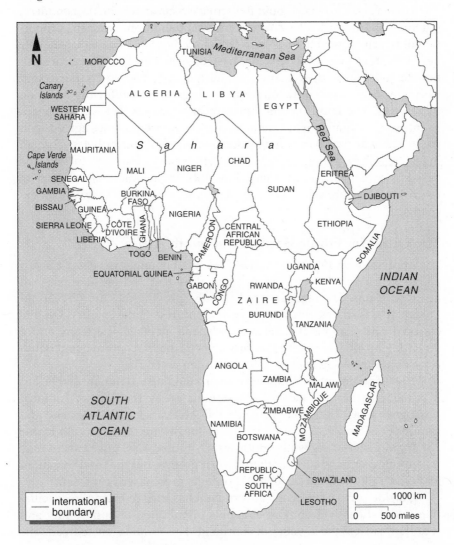

Africa's Search For Stability

Africa remains the world's poorest continent. Of the 30 countries categorised by the World Bank as Severely Indebted Low Income Countries, 25 are in Africa. In some countries, life expectancy is falling and infant mortality is rising. Even so, the overall population is expected to double by 2020, far exceeding economic growth. In Uganda and other countries first hit by the AIDS pandemic, death has removed part of a generation, often killing the young and skilled, leaving the old and children unable to fend for themselves.

There is no consensus about the actual effects of debt on the continent, but Sub-Saharan Africa now owes some $211 billion, an increasing proportion of which is owed to the World Bank and the IMF. The Bank, led by its new President James Wolfensohn, has suggested a comprehensive solution to relieve the debt burden, while the IMF, backed by Japan and Germany, argues that only four or five countries have difficulty repaying their debts. The Fund prefers a piecemeal approach that would benefit countries like Uganda which implement reform programmes agreed with the IMF, but are still hampered by outstanding debt. Nevertheless, foreign investment is beginning to flow into Africa again. From 1988 to 1991, the tide for investment was waning, but since then, funds have begun to return, largely through South Africa, but also to Africa's new stock exchanges.

The Quest for Ethnic Harmony in South Africa

The impression in 1995 that South Africa was booming was given a sharp jolt in February 1996 by rumours that President Nelson Mandela had heart problems. The rand dropped sharply, proving that the 77-year-old President was still essential to peace and progress in the country. Despite his own protestations that he was only one member of a team, and a three-day hospital check-up that proclaimed him fit, concern about South Africa after the Mandela magic has gone remained rife.

The Government of National Unity is mandated to run until 1999, but it is likely to become increasingly dominated by the African National Congress (ANC). Relations between the ANC, including Mandela, and Deputy President F. W. De Klerk are hostile, as are relations with Home Affairs Minister Mangosuthu Buthelezi. A future ANC government may co-opt representatives from other interest groups, but the Government of National Unity will not be renewed once its present writ runs out.

Outside party politics, the government has tried to balance reconciliation with whites with a more equitable distribution of opportunity and wealth to redress the injustice of apartheid. This is a fine line to tread. The 1994 settlement was only an interim one and some essential issues are only now being fought out in the details of the new Constitution scheduled to be announced in May 1996. Cyril Ramaphosa, the ANC General-Secretary and Chair of the commission on the new Constitution, grappled with such fundamental problems as how much autonomy to give Kwa-Zulu Natal. At the 1994 elections, the ANC had virtually 'bequeathed' the province to the Inkatha Freedom Party and Chief Buthelezi. But since the province is almost evenly divided between supporters of Inkatha and the

ANC, this was not a long-term solution. Meanwhile, the killing continued as rival supporters clashed violently in the rural areas.

The issues affecting whites were jobs and land. The 'sunset clauses' in the 1994 settlement convinced the majority of state-employed whites to stay in their jobs. But they are already among the most highly taxed citizens in the world, and feel threatened by attempts to insert 'socio-economic rights' into the Constitution. Success in sporting events has been cheered by all and this is not to be underestimated, but South Africa remains a deeply divided society. The refusal of Afrikaners in Potgietersrus to allow black children into their school in January 1996 was an exception rather than the rule and the issue was quickly resolved. But more serious problems will arise if affirmative action campaigns are organised badly, and if either blacks feel nothing has changed or whites feel they have become the victims of reverse discrimination. The question of land will also soon arise, with sections of the ANC wanting to return land sequestered under apartheid to its original owners, and white farmers demanding limited redistribution and proper compensation.

Building homes, creating jobs and other elements in the Rehabilitation and Development Programme launched with such fanfare when the ANC came to power lagged in 1995, but concern over these matters has still not translated into open frustration. The common view is that as long as blacks can see some development somewhere they will believe that it will, one day, benefit them or their children. What no one can predict is how long this attitude will prevail.

South and southern Africa remain the continent's focal point to the outside world. The belief is that if South Africa fails there is no hope for the rest of Africa, while if South Africa succeeds it just might have a beneficial effect on the continent as a whole, at least on the southern zone. In fact, South Africa's incoherent foreign policy and lack of commitment to the rest of the continent is leading it towards a 'fortress South Africa' stance.

West Africa in Nigeria's Shadow

South Africa's incoherent foreign policy was brought into sharp relief in November 1995 when Nigeria hung Ken Saro-Wiwa, a writer and campaigner for the Ogoni people, and eight of his colleagues. The timing, right in the middle of the Commonwealth Conference in Auckland, New Zealand, was perceived as a calculated snub to the community. But it is more likely that the strong belief among world leaders that Nigeria's President General Sani Abacha would not carry out the death sentences was matched by his ignorance of the international outrage his action would provoke. At the time, African leaders at the Conference, including Chief Emeka Anyaoku, the Nigerian Commonwealth Secretary-General, and Nelson Mandela, were reassuring the outside world that dialogue with the Abacha regime would save Saro-Wiwa, just as it had saved the alleged coup plotters – including former head of state General Olusegan Obasanko and his former deputy General Shehu Yar'adua – in March 1995.

From his bastion in the capital, Abuja, General Abacha defied the world while Nigerians faced shortages, rising prices and plummeting health and education services. Fears that Nigeria will explode into civil war, however, are probably exaggerated. Although from the outside it may seem that politically and economically Nigeria is travelling rapidly towards the buffers, this may be a false image. Instead, disorder may dissipate rather than escalate into rebellion, and the government may break up into rival camps or disintegrate rather than end in confrontation. The widespread lack of faith in politics and politicians in Nigeria makes for poor politics, but also prevents the build-up of strong political movements. Yet as long as the government receives some $2 million a day from oil revenues, it will be able to buy guns and manipulate opponents. The regime can only be removed by a coup from the inside, which would not necessarily bring about a better government.

In West Africa, Nigeria's position is contradictory. While the United States and other industrialised countries are trying to isolate it and force a return to civilian rule, these countries also support the Nigerian military's peacekeeping roles in Sierra Leone and Liberia. In the former, the Nigerian military is maintaining in power a military government committed to handing over to civilian rule. In Liberia, Nigerian forces held the capital, Monrovia, kept out Charles Taylor and his forces and looted quantities of moveable goods. Only when Taylor met General Abacha at Abuja in August 1995 was a deal struck. Taylor seems to have persuaded Abacha that he could become president of Liberia after all, and Nigeria's troops have now stopped attacking him.

This ambivalence in Nigeria's role in West Africa is reflected in the struggle between civilian and military governments in the region. It was alarming – if not surprising – to hear Lieutenant-Colonel Ibrahim Barre Mainassara, who took power in Niger in January 1996, praising General Abacha, saying that Africa needed more leaders like him. Nigeria's vast oil incomes give it immense weight in the region. It is perhaps significant that Valentine Strasser, Sierra Leone's young military head of state who voted against Nigeria at the November 1995 Commonwealth meeting, was overthrown on 16 January 1996. While Sierra Leone went to the polls in February 1996 to elect a civilian government to replace the military regime that had ruled for 11 years, there was widespread doubt that the election could solve either the civil war or the problem of large numbers of young people with guns and little political purpose beyond making a living from them.

In Gambia, international pressure has had little success in forcing the young soldiers who seized power in 1994 back into the barracks. This was particularly galling for the UK, which has centuries of close ties with Gambia and had presented it as a shining example of multi-party democracy in Africa. Coup attempts in Guinea and Congo in early 1996 suggest that military rule, far from being banished by the democratic movement of the early 1990s, may be making a comeback. On the other hand, the coup-makers in Niger justified their successful seizure of power, with some merit, as ending the political deadlock in the country that had prevented its civil service and army from being paid. Similarly, the military unrest in Congo could be explained by the fact that soldiers had not been paid, an indirect result of the 50% devaluation of the CFA franc in 1994.

Elections, like coups, can fail to stabilise political life. Côte d'Ivoire, one of the success stories in Africa until recently and still very close to France, disappointed democrats with a deeply flawed election in October 1995 won by Henri Konan Bedie, with support from France. His main opponent, Alassane Ouattara, was disqualified on a technicality and other challengers withdrew, leaving a one-sided contest and an unconvincing victory.

In 1996, the most important test of democracy will be in Ghana, where Jerry Rawlings, who transformed himself from a young military dictator to a democratically elected leader in 1992, faces an election in November 1996. Ghana has been described by Western donors and the World Bank as a model of economic success after a 5% average growth rate in the early 1990s. Its economic policies have benefited cocoa farmers and mining companies, and earned millions of dollars of aid. But politics could again ruin Ghana and the omens are not good. In December 1995 Rawlings, an unpredictable and volatile character, assaulted his vice-president at a Cabinet meeting. If the election is free and fair, which is doubtful, and Rawlings loses, he is unlikely to step down quietly. Democratic failure in Ghana would leave only Benin and Mali, two lightweight and vulnerable states, as West Africa's democratic success stories.

Whither the State?

But this concentration on parties and elections may obscure the deeper political dynamic in West Africa. A new phenomenon, starkest in Liberia and Sierra Leone, may sweep away such fragile institutions as political parties, elections and possibly even states and borders. As central government weakens, under pressure from donors and local or ethnic solidarity, new power centres are emerging. Some of these are old institutions, such as chiefdoms or kingdoms, or old trading networks. Others are new, such as the armed gangs who roam the countryside living off the land, or the local defence committees set up to defend the people against them.

This phenomenon is graphically illustrated by the emergence of Executive Outcomes (EO). This is a registered company of mercenaries, mainly ex-South African special forces, who acted as apartheid's cross-border raiders and death squads inside the country. They first came to notice in Angola where they were employed by the *Movimento Popular para a Libertação de Angola* (MPLA) government to clear UNITA rebels out of the Soya oilfields. Many of the EO fighters had previously fought alongside UNITA in southern Angola, but had now changed sides. They were so successful that the MPLA then employed them to use fighter bombers and gunships to help drive UNITA out of its main bases in the central highlands.

In Angola, EO was paid in dollars, $40m a year. But its high-profile success there led other countries, particularly Uganda and Sierra Leone, to seek its services also. In Sierra Leone, however, the government had no way of paying the mercenaries except to grant the company a diamond mining concession. In exchange, EO troops flew gunships and helped track down and kill rebels of the Revolutionary United Front. By acquiring real estate, however, EO was no longer a hired hand. It had established a stake in the country and must stay there for some time to recoup profits.

EO may look like the last throw of a failing government which can no longer rely on its own troops to defend it, but the phenomenon has deeper implications. As African states weaken, it will not only be the local barons who take over. Africa's mineral wealth attracts outsiders who, unable to rely on a government army to protect their operations, will employ companies like EO. That, in turn, will further weaken central government and even threaten to dismember African states.

It seems unlikely that countries like Sierra Leone and Liberia will ever be nation-states again, as they once were in the 1960s and 1970s. At best, the 'central' government in the capital will be forced to come to an agreement, spoken or unspoken, with the new powers in the rest of the country leading to a separation of territory or powers or both. In these countries this settlement has come about through war and disintegration, but a similar process is happening elsewhere through less violent means.

Governments are not only under pressure to privatise state companies that have been an important source of patronage and power, and to reduce government control of their economies. They are also being encouraged by circumstances, as well as external pressure, to form regional economic groupings. Last year the East African Community, moribund since the early 1970s, was revived and commitments, as yet unfulfilled, were made to remove tariff barriers between Kenya, Tanzania and Uganda. The election of Ben Mkapa in Tanzania in October 1995 has helped this process; Kenya's reservations and incipient mistrust of Uganda were, at least temporarily, overcome. Similar moves are afoot in southern Africa, where the old South African customs union is merging with the Southern African Development Community, formerly the Southern African Development Coordination Conference, the front-line states' attempt to reduce their dependence on South African budgets. The net effect of these changes is to reduce the powers of central governments and weaken states, thus creating power vacuums to be filled by local authority, legitimate or otherwise.

Two countries actively encouraging such changes are Ethiopia and Uganda. In Ethiopia, Prime Minister Meles Zenawi is trying to create a new political order based on ethnicity. His aim is to reduce the role of central government in local affairs, leaving it to local assemblies and regional, ethnically based administrations to build roads and maintain schools and hospitals. Centralised power has been at the root of Ethiopia's problems for centuries and devolution there is seen as the answer.

This change also suits President Meles, who rules through a coalition of ethnically based parties dominated by his own Tigrayan group. Since they have neither sufficient numbers nor resources they cannot hope to dominate Ethiopia alone, but if they can retain the essentials of government, including the army, they can allow every other group to run its own area. They can even, in theory, allow regions to secede, as Eritrea has done. Whether that would be allowed in practice is another matter and any political party that has challenged Meles by not accepting his broad plan for Ethiopia has been crushed and its leaders imprisoned.

Devolution is also taking place in Uganda where the government is encouraging towns and regions to look to their own resources instead of expecting hand-outs from the centre. But in Uganda there is a real danger of atavistic elements as Buganda, the old kingdom, tries to secure its autonomy or even secession from the rest of the country. President Yoweri Museveni, who came to power 10 years ago at the head of a peasant army, faces an election later this year. He has argued that religion and ethnicity have wrecked Uganda in the past and banned political parties from campaigning, although everyone knows which political party each leader represents. If Paul Ssemogerere, a Catholic of the Muganda group and Museveni's main rival, manages to canvas these two constituencies and defeat or run Museveni close, Uganda could again become unstable.

Weak Links in the Peace

Uganda may also become embroiled in the Great Lakes apocalypse. Yoweri Museveni, whose own Ankole group is divided into the ethnic equivalent of Hutu and Tutsi, supported the Rwandan Patriotic Front (RPF) in its struggle. Though he may not have encouraged their war in the first place, most of the RPF officers were Rwandan Tutsi who had grown up in exile in Uganda and had been senior commanders in Museveni's own National Resistance Movement. There has, however, been no political progress in the Great Lakes, and another massacre could take place at any time. Meanwhile, in Burundi, steady attrition by murder is having the same result as in Rwanda, only more slowly. Despite threats by Zaire's President Mobutu Sese Seko to send the Rwandan Hutu refugees back to Rwanda, he has not acted on them and is unlikely to. Western support for his continued rule depends on some stability in Zaire and though at present only he can provide that, he still needs to be mindful of Western concerns.

The failure of the local countries bordering the Great Lakes to coordinate their efforts to bring peace to the region demonstrates that Africa is not capable, or willing, to solve its own problems. The Organisation of African Unity (OAU) remains a broken reed and only former US President Jimmy Carter was able to bring together the regional heads of state to discuss the problem. The weakness of the United Nations and its virtual expulsion from Rwanda suggests that there will not be a decisive or strong response if the killing and mass exodus begin again. The situation in Angola, too, has highlighted that outsiders cannot impose solutions in Africa. In Mozambique, an agreement held because both sides stuck to it. In Angola, the weak and depleted UN could do little when UNITA broke, and continues to break, the peace accord, and the government compounds this breakdown. It now seems likely that UNITA will not only have a place in central government, but will also retain some territorial control, possibly being offered a diamond mine as a source of income and retaining some of its fighting capacity.

Another shortfall in democracy was registered in October 1995 by Tanzania, one of the last countries on the continent to hold a multi-party election. In the end, Ben Mkapa, the candidate of the *Chama Cha Mapinduzi*, which has ruled Tanzania as a one-party state since independence, won. But the election was surprisingly chaotic, given that Tanzania had previously held efficient elections under the one-party system. In Pemba, the heartland of the Civic United Front (CUF) which wants more autonomy for the islands, the election was marred by fraud. Too many people doubt the truth of the results and unless a deal can be struck to bring the CUF leader, Sharrif Hamad, into government, or at least grant some Zanzibari and Pemba demands, there could be trouble in the future.

Mkapa's reputation rests on clean hands, uncommon in Tanzania. But whether he will be strong enough to abolish the corruption that has spread throughout the country remains to be seen. One of his first acts was to restore links with the IMF and other aid donors, which may lay the foundations for long-term reform, but in the short term may exacerbate politically dangerous unemployment and rising prices.

Kenyan President Daniel arap Moi has complained openly that multi-party democracy was forced on him and that he does not believe in it. He acted accordingly, refusing to register Safina, a small party which hoped to coordinate the opposition for elections in 1997. Its leaders, including the prominent white Kenyan Richard Leakey, have been beaten up in public and derided by Moi.

Faced with this and the aftermath of the huge Goldenberg corruption scandal in which over one hundred million dollars were stolen from Kenya's state coffers by its top politicians, Western donors came near to cutting off aid to Kenya, but drew back as usual at the last minute and agreed to continue their support. The threats and counter-threats between the Kenyan government and the donors have become an annual ritual dance. It usually ends with Kenya promising to clean up corruption and cut spending and the donors, needing a stable Kenya as a base in the region, paying up.

Aggression in East Africa

Last year there was a concerted effort, backed by the United States, to isolate Sudan. The US accuses the Islamist government, headed by General Omar Beshir but controlled by Hassan el Turabi, who preaches fundamentalist Islam, of supporting terrorism. The regime has few friends. The Sudan-backed attempt to assassinate Egyptian President Hosni Mubarak in Addis Ababa in June 1995 led to Sudan's unprecedented condemnation by all governments in the OAU and by most in the UN. Ethiopia launched a diplomatic offensive, demanding the return of the suspect assassins while Eritrea went further. President Issiais Afwerke gave safe haven to Sudanese rebels pledged to overthrow the Khartoum government by force and publicly stated his support and his willingness to supply guns.

Khartoum's isolation also helped the Sudan People's Liberation Army (SPLA), the southern movement that rejects northern Arabic and Muslim domination and seeks greater autonomy, and perhaps independence. Having set aside its internecine tribal war, the SPLA has re-armed and has retaken much of the territory lost during its disastrous break-up in 1992–93. In late July and early August 1995, the SPLA launched raids as far as Juba, the capital of the south. This time the movement seemed determined to take the city and then declare some form of independence for the region. Although this failed, they re-established themselves as a fighting force.

Sudan struck back at its neighbours, suspecting that they were at least providing a channel for arms to the SPLA. In northern Uganda, the Sudanese-backed Lords Resistance Army caused havoc, killing almost 300 in an ambush in January 1996. This has led to a vast increase in Uganda's defence budget, costs the country can ill afford at this time.

Eritrea also pursued a war with Yemen over the Hanish islands in the Red Sea, where fish, oil and tourist potential, rather than the islands themselves, were the prize. This is the first time that an African country has launched an attack outside the continent, demonstrating that despite being poor and weak, Eritrea is ready to use its huge and now unemployed army to pursue territorial aims aggressively.

Repeating History

Democracy has been popular in Africa. Turnout at elections remains high as people want to take part. But the politicians thrown up by elections and the political systems in which democracy is forced to operate in Africa do not seem to be solving the continent's problems. None of this year's elections are likely to bring about real change in African politics. Most will be manipulated by rulers to continue their repressive and exclusive rule. The power of few of these rulers can be manipulated by outside sanction. Equally, the winners in democratic countries will find it difficult to bring about significant economic change because governments are restricted in the policies they can adopt by their poverty and dependency on donors. Donors will not help without a deal with the IMF and the World Bank, but such deals mean floating the currency, privatisation, budget cuts and other measures which reduce the power of governments.

Donors do not set out to weaken governments, but they do have great difficulty in strengthening them. In some countries, such as Kenya and Zaire, non-salubrious leaders are sustained by aid despite the best intentions of the donors, who fear the anarchy that will ensue without an effective leader at the head. In other countries, the rules that surround the aid make it difficult for the weak leader to follow; here donors have difficulty influencing, and the governments have difficulty implementing, positive change. In other countries, governments are so weak that aid is stymied. Nation-building through aid has always been a losing proposition.

The weakening of the state in Africa is causing deeper, unforeseen changes. Power is slipping away from the towns to the rural and urban periphery. It may be significant that in Uganda and Ethiopia, the two most radical alternatives to traditional political systems in Africa, the new leaders came to power at the head of peasant armies and both had lived and fought for several years in the bush.

Liberia and Sierra Leone are both victims of rural rebellions that have disintegrated into banditry. These countries will not have peace unless those movements are brought into the political mainstream and their thousands of armed young men are given the prospect of education and employment. In the present climate, where aid to Africa is likely to be sharply reduced in the next few years, such improvement will not come about quickly. That is not to say that there will be no economic growth in the continent. War and weak states offer new opportunities both for outsiders and local opportunists, but the pattern which may emerge may recall the pre-colonial symbiotic relationship that existed between charter companies or merchant traders and local chiefs and kings. Those kings and chiefs are already emerging away from the capitals and central governments. In Africa, barons are replacing the bureaucrats.

Rwanda And Burundi Still Threatened

The ethnic aggression and communal malice demonstrated by the ancient Christian and Muslim societies in the Balkans in recent years outraged world opinion. The horrific slaughter in Rwanda two years ago, however, like Cambodia's 'Year Zero', produced a different, more despairing reaction: not outrage, but horror and incomprehension. Passionate debate among democracies about how (or whether) to restrain ethnic homicide in Bosnia finally produced NATO air-strikes and the Dayton accord. The world, however, had no such answers for the tragedy in Rwanda, and was reminded again how impotent power can be in the face of anarchy that turns even members of religious orders into assassins and pits against each other populations at near-subsistence levels with so little of value to defend. One year later, the world continues to wait and watch as the drama unfolding between the Hutu and Tutsi shifts from Rwanda to neighbouring Burundi.

Rwanda: Fragile Truce But No Reconciliation

In 1995, 18 months after the start of the genocide campaign waged by the former Rwandan government, the country began to return to normal. Since a new broad-based government was established in July 1994, the humanitarian situation in Rwanda has steadily eased, but this is only encouraging when compared with the catastrophic breakdown of society in previous years. While the overall situation in Rwanda remained calm in 1995, 'calm' is a relative concept: stability in the western prefectures continues to be threatened by the former government's military and militia, while banditry, sabotage, assassinations and intimidation of local populations have steadily increased, particularly in border areas. Continuing human-rights abuses also remain of concern.

Outside a relatively bustling capital, many thousands of Rwandans are without homes, land or livelihoods. In a country that has always been almost entirely dependent on subsistence farming, food production is now only 60% of normal levels and the production of principal export crops is only 40% of previous harvests. The government has set itself the clear task of moving from humanitarian assistance to rehabilitation, and then reconstruction and development, but the task will be daunting.

The Challenge of Repatriation

Repatriation of Hutu populations which had begun to flee before the Rwandan Patriotic Front victory in July 1994, remains a Damoclean sword over the country's fragile peace. Approximately 1.7 million Hutu are presently quartered in Zaire alone, and more than 700,000 refugees fled Rwanda last year. While the UN has made contingency plans for a possible large-scale return of 300,000 people, the problems involved in reabsorbing a massive or uncontrolled return of refugees would, undoubtedly, have a serious impact on the country's stability.

Fear is a powerful factor preventing the return of many who have fled, but this hesitation is interpreted as a sign of guilt by many Tutsi who survived the massacres. Not least among the inducements for Hutu repatriation, therefore, must be a greater level of assurance that returning refugees will be protected from revenge attacks or arrest by those who suffered at the hands of the previous government. The prospect of arrest and imprisonment, and the terrible conditions existing in Rwanda's overcrowded jails, is another consideration which discourages many Hutu from returning home.

By the beginning of 1996, expanded prison and detention space and better health services had led to some improvement. The rate of arrests remains high, however, and with some 65,000 people already incarcerated, creating new detention facilities cannot alone keep pace with demand. The government has repeatedly stressed that all Rwandans have the right to come home, and it has developed an accelerated plan of action for 1996, targeting the communes with the greatest resettlement problems.

The Hutu responsible for the genocide, meanwhile, are also to some extent responsible for the refugee crisis, having encouraged their fellow tribesmen to take flight before the RPF. They calculated that the international community would respond to a massive tidal wave of refugees, and they were proved right. It was only when hundreds of thousands of Rwandans fled to Tanzania, and then nearly one million more to Zaire, that the international community mobilised its resources. In addition, many of the refugees in neighbouring states remain unreconciled to the new Tutsi-dominated forces and are biding their time in exile – perhaps acquiring arms and training – until the day they hope to return to recapture the government.

Repatriation is also a concern for those countries that shelter the refugees. Rwandan refugees in Burundi were told in January 1996 by the Burundi Army Chief of Staff that they should go home as quickly as possible. Zaire's threat of forced repatriation to Rwanda after a December 1995 deadline for voluntary return focused political attention on the issue. President Mobutu Sese Seko has since withdrawn the threat, but in February 1996 he did use troops in what appeared as much a half-hearted attempt to pressure the donor states as it was to intimidate Rwandans into leaving. Despite the aggressive repatriation campaign by the United Nations High Commission for Refugees (UNHCR) since October 1995, the rate of voluntary return remains very low. During September and October 1995, a total of 32,190 refugees returned to Rwanda, mainly in UNHCR-organised convoys. The numbers returning voluntarily from Zaire decreased in November, and in Tanzania and Burundi virtually came to a halt.

The events of 1994 left Rwanda and its population severely traumatised. Millions are dispossessed or displaced; and more than 40,000 children are without parents or are separated from their families. Many have either been the victims of brutality, or witnesses to the horror of genocide. Much of the country's infrastructure has been devastated, its legal system has broken down, and its pool of skilled professionals depleted as a result of genocide, massacres and refugee emigration. This legacy will continue to weigh heavily on Rwandan society, and will compound the fear that makes many survivors reluctant to pick up the thread of their former lives in their home communities.

Before the country – its people and its leaders – lies the major task of restoring the fabric of civil society, of rebuilding its infrastructure, rehabilitating the economy and providing justice without revenge. It must do so in the face of a massive refugee problem and a relatively poor economy which has been gravely disrupted. And all this must be accomplished while recognising that the same ethnic passions that have already brought Rwanda itself to near ruin exist in its immediate neighbour, Burundi.

Burundi: Awaiting Disaster

In 1996, world attention – and apprehension – has come to focus on Burundi. Like Rwanda, Burundi is a backward, agricultural society in which Hutu tribesmen dominate in numbers (about 80–85% of the population), but where the minority Tutsi have historically controlled the military and dominated political life. The Hutu have the numbers, as one observer pointed out, but the Tutsi have the guns. Moreover, the Tutsi used them in

1972 to crush a Hutu uprising causing some 100,000 Hutu casualties. Here, too, the same explosive mixture of ethnic fear and hatred, so far barely controlled, may yet cause a tragic re-run of the catastrophe that befell Rwanda a few years ago.

The present crisis stems from a turning-point in modern Burundian history – the July 1993 election of Melchior Ndadaye, a moderate Hutu, as President, whose margin of some 65% of the vote was both a victory for democratic process and an important step towards providing the majority Hutu with a larger share of political influence. The democratic honeymoon period was not to last long. On 21 October, the 11th Parachute Battalion seized the Presidential Palace and executed Ndadaye and a number of his staff.

Although this *coup d'état* was immediately quelled, it triggered reprisal attacks against Tutsi populations in rural areas and precipitated ethnic conflict resulting in the deaths of up to 100,000 Tutsi and Hutu. Over 280,000 people were displaced throughout most of the central and northern provinces of the country and nearly 670,000 refugees fled to Tanzania, Rwanda and Zaire. By July 1994, the advance of the Tutsi RPF through southern Rwanda triggered a further spontaneous return of Burundian Hutu refugees to Burundi. A period of tension, fear and episodic communal killing began. To all practical purposes ungovernable, the country has remained on the brink of outright communal warfare ever since, with a Rwandan-style genocide kept at bay only by a balance of terror between the majority Hutu and the armed Tutsi. The Hutu-dominated Front for Democracy in Burundi and the Tutsi Union for National Progress are locked into sharing a government which they would each prefer to deny the other.

The genocide in Rwanda has encouraged extremist Tutsi ruthlessness towards Hutu neighbours, and gangs of armed Tutsi have 'ethnically cleansed' Bujumbura, driving the Hutu into ghetto-like suburbs or into the countryside where they have operated as marauding bands. Hutu extremists, in turn, are reinforced by the *Interahamwe*, Rwanda's former Hutu Army and armed militia. Given the success of the RPF in Rwanda, Tutsi have asked why they should not take advantage of this historic opportunity to control both countries.

Against this background, international efforts at conflict mediation and preventive diplomacy have struggled to keep the country from exploding. Relief programmes were mounted after Ndadaye's assassination to address the special needs of both groups, led principally by the International Committee of the Red Cross (ICRC), beginning with emergency medical care for the wounded. Shortly thereafter, food aid programmes were organised to assist the displaced who had been unable to grow their own food after missing the planting season. The number of food-aid recipients grew rapidly, reaching 800,000 in May 1994. In response to the developing crisis, the number of humanitarian agencies in Burundi expanded from about ten to over 30.

The increasing instability of the country is reflected in the changing nature of the foreign help it has received. In 1992, such aid stood at a high of $322 million, of which $223.9m represented investment and aid programmes and $94.3m technical assistance, while emergency and humanitarian aid accounted for only $3.7m. By 1994, the figures were even more dramatically reversed: out of total assistance of $220m, only $65m constituted investment and aid programmes and $60m technical assistance, while emergency aid accounted for $95m. Preliminary figures for 1995 indicate a further drop in overall investment as the country's prospects for economic development remain on hold.

Throughout 1995, clashes between armed Hutu militia in the west, using Zaire as a safe haven, as well as in Bujumbura itself, resulted in many casualties, the destruction of property and population displacement. Conflict also spread to the east and south. The province of Cibitoke was declared a war zone in April 1995 and fighting also broke out in the neighbouring province of Bubanza, where more than 100,000 displaced people were reported to be in urgent need of humanitarian assistance.

Since October 1993, Burundi has been enmeshed in a continuing political crisis, with Tutsi political parties attempting to regain control of the governmental apparatus and competing among themselves for influence. The government, however, lacks any real power and is not in a position to disarm either side. The absence of the rule of law in Burundi and the near paralysis of the government continue to undermine the existing administrative structures and depress public confidence. The government's ability to deal effectively with this situation remains, at best, questionable.

Burundians speak frankly of their lack of confidence in a government whose members are closely associated with past massacres. Some of the men guilty of the events that precipitated Burundi's descent into near anarchy not only seem beyond justice, but remain important players in the political game. This compromises the state in the eyes of the public. In August 1995, the UN Security Council established a commission to investigate the events of Ndadaye's death and subsequent killings with a view to identifying the guilty. But in January 1996, UN Secretary-General Boutros Boutros-Ghali wrote to the Security Council informing them that the many problems on the ground made it unlikely that the commission would report in the near future. Such developments confirm the belief that extremism can be practised with impunity. Until this issue is addressed satisfactorily, Burundi can expect further violence.

Nor can Burundi's senior leadership claim to have done much to win back public confidence. Neither side is interested in compromise and conciliation, although both sides recognise that any other road will lead to disaster. Despite the efforts of the relief community to encourage politicians to act constructively, they have gradually distanced themselves from the business of government, and moderate elements from both sides have fled the country, fearful for their safety. Those in the capital are consumed by political infighting, leaving provincial administrators feeling that they have been abandoned by the central government.

The last two years of political instability and conflict have inevitably undermined Burundi's economy, further compromising the country's ability to emerge from its crisis. Burundi's agriculture contributes over 45% of the country's annual average gross domestic product (GDP) of more than $1 billion, and employs 90% of the population. Since the present crisis began in 1993, agricultural production has fallen by 22%, necessitating substantial food imports for distressed populations. Preliminary indications show that this year's harvest of traditional subsistence crops will be close to average, but insecurity may drive many farmers from their land, leading to dramatically reduced food availability in 1996.

The rising level of indebtedness further darkens Burundi's economic prospects. Taking 1990 as the baseline with an index of 100, indebtedness had risen to over 130 by 1995. Perhaps the greatest impact on the economy has been the departure of private business and investment, with the result that humanitarian assistance and its associated services have become a major source of business and employment. It seems likely that the trend of economic donor withdrawal from Burundi will continue in 1996.

Humanitarian Intervention

The main focus of relief-aid programmes in Burundi has been to provide assistance to the Rwandan refugees who have settled there and to returnees repatriating from Zaire. These humanitarian operations in Burundi take place in a very precarious security situation: threats and attacks against relief workers have been common for some time and ambushes have restricted travel by road to the interior. The murder of at least ten aid workers in Burundi in 1995 dramatically underlines the risks of providing assistance in such an ethnically tense environment. It has become apparent that within the polarised body politic, humanitarian agencies have increasingly become pawns in a game of political propaganda being played by both sides. Organisations working predominantly with Rwandan refugees, or in rural areas with dispersed populations, are accused of sympathy towards the Hutu, while organisations working mainly in the towns are denounced as favouring the Tutsi.

Towards the end of 1995, following grenade attacks on relief agencies in Gitega and Ngozi, some international emergency operations were halted, and the ICRC took the unusual step of stopping operations and withdrawing most of its personnel. The departure of all but UNHCR personnel from the north of the country has greatly affected programmes for displaced persons and refugees who depend on aid for their survival.

Little Hope for the Future

On 16 February 1996, in his report to the Security Council, the Secretary-General once again stressed his concern over Burundi. He asked the Security Council to consider a plan to station a rapid-reaction force in a nearby country in case the security situation deteriorated dramatically. This was rejected. The Security Council decided instead on a three-track approach:

- to support the domestic political process by encouraging all parties to work for reconciliation;
- to prepare contingency arrangements to protect the local population; and
- to encourage mediation efforts by regional leaders and the OAU.

Although many of the ingredients for disaster are present in Burundi, demographic factors suggest that Rwandan-style genocide may not be repeated there. Hutu and Tutsi populations are not mixed in the areas of greatest insecurity in Burundi, and intensifying the civil conflict in the north of the country, although creating large-scale population displacement, would probably not provoke the levels of killing witnessed in Rwanda between April and July 1994. Fighting in the central and southern regions of Burundi, where ethnic mixture is more prevalent, could result in a significant number of deaths within the two communities, although here, too, massive displacement rather than genocide is more likely.

The question still remains of how long the armed, low-intensity violence can continue before erupting into full-scale communal warfare. The suggestion by Boutros Boutros-Ghali of deploying a rapid-reaction force to prevent mass killing has not been considered by the major powers because no state is prepared to contribute the requisite forces. Yet the international community may have to face human distress on a scale that it has come to regard as unacceptable and – even worse – may have to accept yet another episode of genocide, which it has solemnly and formally pronounced would not be tolerated in today's world.

Thus stated, the problem is urgent, but the solution is unlikely to come from outside forces. It is the political leadership of the countries themselves that ultimately holds in its hands any real prospect of breaking the cycle of violence and revenge, but this leadership has demonstrated that it is more a part of the problem than a likely source of solution. There remains, therefore, only the course of trying to isolate extremists and shore up moderates while hoping that the embers of ethnic hatred and revenge eventually burn themselves out.

Nigeria: A Nation In Crisis

The hanging of the writer Ken Saro-Wiwa and eight other minority activists from the oil-rich Ogoniland in south-eastern Nigeria on 10 November 1995 brought home the grim reality that another potential threat to international peace was taking shape in Africa's most populous country. Although it had taken the sheer insensitivity of the executions to focus world attention on the precarious state of the Nigerian nation, the crisis has been building for the past three years as an increasingly autocratic military regime has been steadily abusing its power.

Background to Crisis

On 12 June 1993, Nigerians went to the polls to elect a new president after nine years of uninterrupted military rule. The election, against all odds, was the most peaceful and the fairest ever held in the country and Chief Moshood K. O. Abiola, the candidate of the Social Democratic Party, was widely acknowledged to have won. For the military, however, the wrong man had triumphed. On 23 June 1993, Nigeria's then military leader, General Ibrahim Badamasi Babangida, annulled the election, claiming that the process leading up to it had been flawed.

Inevitably, this infuriated a normally restrained population, many of whom had seen the election not only as a victory for the Social Democratic Party, but as a vote against continued military rule. In response, the Campaign for Democracy, a coalition of human-rights and labour organisations, called for nation-wide demonstrations. This pressure led to General Babangida's reluctant departure from office on 26 August 1993. Power, however, did not pass to the man who had won the election, but to a quasi-military interim government under Chief Ernest Shonekan, a businessman and close friend of the military.

On 17 November 1993, the military's blatant return to power, under General Babangida's erstwhile Defence Minister and *de facto* deputy, General Sani Abacha, exacerbated the situation. Since then, the new regime has been virtually paralysed, the political economy 'adrift, divided and indebted' in the words of one commentator, and society on the verge of internal strife. The desire for democratisation is not the only victim. A palpable fear of ethnic repression reminiscent of the period prior to the 1967–70 civil war has resurfaced.

Some believe that it was Saro-Wiwa's challenge to such repression, in an oil-rich area, that led the government to impose the death penalty, hoping no doubt that it would serve as a deterrent to any other disaffected minority community in the Niger Delta that might consider a similar path. From his bastion in the capital, Abuja, General Abacha defies the world while Nigerians face shortages, rising prices and plummeting health services. Defensive, stubborn and aloof, General Abacha was further isolated by the death in a plane crash in January 1996 of his eldest son, Ibrahim, who had served as his emissary and business manager.

While the proximate causes of Nigeria's unrelenting crisis can be traced back to the still-born 1993 elections, the roots of the crisis reach all the way back to the birth of the country and its early development. Nigeria's faulty federal structure led to the military's first foray into politics in January 1966, and for 25 of its 35 years of independence from the UK Nigeria has been subject to military rule. Civilian politicians, although in office for less than a decade – and then often as a front for the military – must take their share of responsibility for the country's failure to sustain democratic institutions. Nonetheless, the major share of that responsibility falls on a succession of predatory military regimes which alienated even those who might once have defended the military as a force for national unity. By now, the longevity of military rule in Nigeria has turned it into the only real centre of power, able to manipulate, and so far to contain, the divisive forces of ethnicity, religion and regionalism in Nigeria's body politic.

The political tensions following Nigeria's independence on 1 October 1960 stemmed from the new state's regional structure and the uneven development of its constituent regions. Although three major ethnic groups – Hausa-Fulani, Yoruba and Ibo – dominate Nigeria, the country is actually home to no fewer than 250 ethnic groups, such as the Ogoni of the Niger Delta. All have different ideas of what the nation they now belong to should be. These inherent, structural divisions operate against national integration; superimposing on them a 'winner-takes-all' political culture is a prescription for chaos and violence.

Such structural problems, moreover, are compounded in a society where political patronage, official corruption and profligacy have become the rule. Thus, control of the political centre is vital because it is the central government that dominates the vast oil revenues and maintains the relative underdevelopment of other sectors of society. Given the struggle of the elites to win this control over oil wealth, politics has become a life or death matter in Nigeria. Successive military rulers have relied on the legal and illegal distribution of public resources to buy compliance and loyalty.

Tightening the Grip

When he came to power in 1993, Abacha adopted the characteristic rhetoric of his predecessors – promises to 'clean house', free the nation's economy from corruption and ruin, reduce dependence on the fluctuating international market and return the economy to the ordinary Nigerians. It soon became apparent, however, that the regime lacked any clear policy direction and saw mere high-handedness as the mark of efficiency. No sooner had General Abacha consolidated himself in power by preaching 'local solutions to local problems', than he embraced the IMF restructuring programme that he had originally blamed for the country's economic woes.

Oil remains central to Nigeria's economic development, accounting for more than 90% of its exports, 25% of its GDP and 80% of its public revenues. Nigeria's access to export earnings from oil sales should ensure its citizenry a reasonable standard of living. Indeed, according to a recent World Bank study, a small increase in oil price can have a major impact on the economy: a $1-per-barrel increase in the oil price in the early 1990s, for example, increased Nigeria's foreign exchange earnings by about $650 million (2% of GDP). Instead of investing this revenue in economic development and programmes for social benefit, however, the regime has used it to divide the population, impoverish its citizenry and undermine democratic accountability. Mismanagement and corruption on a pharaonic scale under this regime have ensured that the proceeds of oil sales have ended up in the pockets of the ruling clique.

General Abacha has only been following the example of previous military regimes, albeit with more vigour. In 1993, according to the World Bank, Nigeria was among the 20 poorest countries in the world, and the situation has since worsened. Gross national product (GNP) grew by only 2.9% in 1994, inflation is currently running at over 75%, external debt has ballooned to $37 billion, unemployment has soared and the Nigerian naira has virtually collapsed. As one recent study notes, 'virtually all pretence of professional economic management has been abandoned, and the government has cynically allowed the economy to become completely predatory in nature'. As a result, the country has stopped servicing interest payments on much of its more than $35bn foreign debt, and it is over $7bn in arrears on its debt to the Paris Club of Western creditors alone.

Although the return of the military was widely viewed with scepticism by Nigeria's civilian population, official propaganda portrayed it as the 'only alternative' to save the nation from disintegration. On the political front, General Abacha won some respect by deftly assembling a broad-based civilian coalition of prominent politicians. His promise to relinquish office soon, his call for a National Constitutional Conference 'with full constituent powers', and his reversal of unpopular economic policies convinced a battle-weary constituency that he should be given the benefit of the doubt. The opportunism and greed of the political class helped. But even in a society atomised by years of military rule and dispirited by cynicism, silence did not necessarily mean acquiescence to military rule.

It soon became apparent that the proposed Constitutional Conference was merely a device to avoid the more all-embracing and sovereign national conference demanded by civil-rights organisations to discuss questions of national unity, federal powers and regional rights. The credibility of the Conference was further undermined by its organisation: the regime would nominate 97 of its members to constitute a quorum in any debate, regardless of the absence of the Conference's 273 elected members. The Conference, moreover, was not to be sovereign nor its decisions binding on the regime.

Largely as a result of an active campaign to boycott the Conference, only about 300,000 voters participated in selecting its representatives, compared to the 14 million who turned out for the presidential election on 12 June 1993. But even with so little reason to expect a constructive and independent gathering, the international community saw it as the only option and was willing to offer it legitimacy. Pressure to have something to show for the investment in it led the Conference to propose a number of genuine changes on thorny issues such as revenue allocations and power rotation. But scepticism about the Conference was vindicated by its endorsement of General Abacha's indefinite stay in power and the stand-offish manner in which its findings and report were treated by the regime. Revelations by a disaffected participant in the 1993 coup that the regime did not in fact intend to limit itself to a brief stay in power strongly supported the view that the Conference was, after all, only part of the government's attempt to create a veneer of legitimacy.

By now the regime was beginning to shed its image of a government with a human face. Human-rights activists were being detained, newspapers arbitrarily proscribed and journalists harassed throughout the country. The populace woke up to the reality that this regime had not come to save democracy as claimed, and calls were renewed for the restoration of civil rights. By May 1994, popular protests were followed by national strikes, including a two-and-a-half-month strike by oil workers which paralysed the industry and had a negative impact on the entire economy. Predictably, the regime retaliated with a fiercely repressive clampdown on the opposition, including the man who won the election, jailed Chief Abiola.

Opposition to the regime was also growing inside General Abacha's military constituency. A military commission set up by the interim government, in which General Abacha was a key figure, concluded that the election results should be released and honoured. The General simply ignored their report. When two of his service chiefs – Major-General Chris Ali, the Chief of Army Staff and Rear-Admiral Alison Madueke, the Chief of Naval Staff – urged the release from detention of Chief Abiola and other pro-democracy activists, General Abacha questioned their loyalty and sacked them. Further repressive measures followed, including the conviction of several retired and serving military officers. Leading democratic activists and journalists, such as former head of state General Olusegun Obasanjo, his former deputy General Shehu Yar'adua and Dr

Beko Ransome Kuti, the leading human-rights campaigner, were prosecuted on the apparently trumped-up charges of plotting to overthrow
the military.

In an effort to combat growing disenchantment within the international community which had so far tolerated the regime, General Abacha
abandoned his half-implemented populist economic programme and re-
verted to the IMF and World Bank prescriptions for devaluation and
liberalisation of the Nigerian economy. This helped to deepen the economic crisis without, however, winning the regime much favour from
these international institutions. The General's three-year transition programme, scheduled to end in 1998, failed to win the confidence of Western
capitals. The execution of Ken Saro-Wiwa and the eight other activists
finally put paid to any hope that the regime would surrender control of its
own accord. The implications of this extend well beyond Nigeria itself.

Impact on Regional Security

Beyond the damage caused to the Nigerian people and the integrity of the
nation, military rule and political crisis in Nigeria have an impact on both
the West African region and the African continent as a whole. Nigeria's
population is roughly equal to that of its 15 West African neighbours, and
is one-quarter of Africa's total population. It is the size of Germany, France
and Greece put together. Its military has always been by far the largest and
best-equipped in the region, and its oil wealth is unmatched in the area.
Nigeria has long been both a source of envy and of pride to most African
countries. Its resources have provided – and in some cases still provide –
the key to prosperity in the region. Put in the simplest terms, if Nigeria
disintegrates, so will regional security.

Moreover, as the military regime tightens its grip on Nigerian civil
society – with little protest from the international community – would-be
dictators in the region may be encouraged to challenge other fledgling
democracies. There is also a risk to others from the refugee crisis that
would follow any serious breakdown of Nigeria's civic society. Nigeria,
which might be expected to exert an influence in favour of democracy in
the region, is now widely believed to be exporting its coup culture to its
neighbours.

Although there is no conclusive evidence, the military coup in Gambia on 22 July 1994 which ousted the longest-standing democracy in the
region, and the 16 January 1996 overthrow of the military leadership in
Sierra Leone, both appear to have been encouraged, if not assisted, by
Nigerian contingents present in both countries at the time. The January
1996 coup in the Niger republic, Nigeria's northern neighbour, was the
second successful *coup d'état* in a month. There is little doubt that the
inspiration for this coup also came from the ruling Nigerian junta. Given

these realities, the Benin republic, Nigeria's neighbour to the west, is taking few chances. President Nicephore Soglo has signed a military understanding with the United States and its military commission recently met in the Benin capital. Nigeria's General Abacha is also known to have close ties with Jerry Rawlings of Ghana, who has pleaded his case in various international forums, and President Gnassingbe Eyadema of Togo, both of them former coup plotters who became civilian presidents under questionable circumstances.

There are other negative security implications for contiguous states. As a result of the porous borders in the area, Nigerians are engaged in money-laundering, drug trafficking and arms smuggling into neighbouring countries. The Nigerian contingent, which formed the bulk of the Economic Community of West African States (ECOWAS) peacekeeping force (ECOMOG) in Liberia, is now seen by some as an army of occupation rather than a peacekeeping force in that traumatised country. Perhaps the most worrying dimension of the drift from instability to chaos in the region is the growing resort to violence as all avenues for legitimate and peaceful protest appear completely closed. Bombs exploded in the northern cities of Kano and Kaduna, as well as in Ilorin, in west central Nigeria, in January 1996 when at least three people were killed and hundreds injured. A hitherto unknown group, the United Front for Nigeria's Liberation, has claimed responsibility for downing the presidential jet in which General Abacha's son and 13 others died in January 1996.

The Role of the International Community

The execution of Ken Saro-Wiwa drew world attention to the barbarity of the Nigerian government and its repressive policies. International sanctions, including those declared but never imposed by the European Union after the annulment of the 1993 election, began to be implemented. In 1995, the spotlight fell on Shell, the company Saro-Wiwa had campaigned against so vigorously in Ogoniland. It soon became clear that while the US and the UK would take certain measures against the regime, they were not willing to impose oil sanctions, perhaps the only measure that would really hurt the Nigerian government.

Despite adopting some measures pioneered by the EU and the US in 1993, and now widened following Nigeria's suspension from the Commonwealth by its heads of government at their meeting in Auckland, the world community still appears unwilling to sanction the Nigerian regime effectively. It is clear that the measures so far adopted are unlikely to have the desired effect of persuading the Nigerian dictatorship to respect human rights and embrace democratic change. This failure can be attributed more to the limited scope of the measures adopted and the uncertainty of their enforcement than to the inherent ineffectiveness of sanctions themselves. The impact of a total arms embargo and tighter visa restrictions, for example, is open to question because it has not been universally applied; only the United States has imposed a ban on its air link with Nigeria, and has been more stringent than other countries in banning travel by regime officials, family members and their collaborators. There is also some evidence that measures adopted since 1993 have not been strictly implemented by certain countries. The UK government, in particular, has shown great reluctance to ensure an effective arms embargo, employing instead a selective definition of the EU embargo and refusing to disclose details of arms deals on grounds of national security.

Nevertheless, evidence suggests that even the limited measures already adopted by the international community are beginning to affect the Nigerian junta's grip on power. If Nigeria's major trading partners, particularly the UK and the US, were to support measures such as the oil embargo and the freezing of assets advocated by the Nordic countries, Germany and France, the impact of sanctions would be even greater. Clashes between the military and civilians in the last three years have seriously affected Nigeria's export earning potential, underlined the regime's pariah status and led to capital flight and disinvestment. Maintaining a front with the help of foreign lobbyists also adds to the cost of maintaining the dictatorship.

Prescription for Change

The current dictatorship in Nigeria and the increased centralisation of power it has required have been maintained by a system of human-rights abuse and a pilfering of state resources –failings that have escaped serious censure from Western powers. Their foreign policy towards Nigeria (to the extent that there is such a thing) is almost entirely determined by the country's huge oil reserves and the market potential of its large population. The West's policy towards Nigeria has lagged behind Nigeria's emergence as the greatest destabilising factor in the region, providing a bad example to anti-democratic forces over its borders. A failure to see the underlying contradiction between progress and authoritarianism has led Western policy towards Nigeria to emphasise the search for 'stability' rather than democratic accountability and justice. Such an ill-advised priority, at the expense of enduring a venal and despotic regime, poses long-term risks not only for Nigeria, but for the entire region.

Clearly, the international community needs to intensify its efforts if it is to prevent a breakdown in Nigerian society that could have far worse implications than those witnessed in Rwanda and Somalia. What Nigeria obviously needs is a broad-based, fixed-term national government. Such a new regime would have the mandate to organise a sovereign national conference which, in turn, could prepare for elections and even reconstitute the army along more professional lines. Insistence on the military's immediate withdrawal from the political arena is a necessary first step: any attempt to move forward without resolving the events of 12 June 1993 would be a total misreading of the special needs of this ethnically diverse state, and an underestimation of the Nigerian people's growing desperation. While the prescription for an improved and stable situation in Nigeria is easy to reach, its implementation is riddled with difficulties.

Chronologies

United States and Canada

January

1 104th US Congress sworn in with the Republican Party controlling both chambers. Newt Gingrich and Bob Dole are elected Speaker of the House and Senate Majority Leader respectively.

24 In his State of the Union address, US President Bill Clinton calls for a smaller government bureaucracy, reduced federal spending and lower taxes, but promises to veto any repeal of gun control.

26 The US House of Representatives approves a Balanced Budget Amendment, 300-132.

February

6 Clinton presents his budget for the 1996 fiscal year, containing limited spending cuts, preserving social security and offering tax benefits for the middle class.

13 In Quebec, the separatist *Bloc Québecois* is defeated by the ruling Liberal Party in two federal by-elections.

March

2 Proposed Balanced Budget Amendment fails to pass US Senate.

6– Canada bans all fishing for two months within 200 miles of its coast in the north-west Atlantic; Canadian patrol boats arrest a Spanish trawler for over-fishing and use of illegal nets in international waters off the coast of Newfoundland (9); EU suspends all formal contacts with Canada (13); Canadian patrol boats force Spanish trawlers to leave disputed waters (26); UK government announces that it will veto any EU trade sanctions against Canada (28).

29 US House of Representatives rejects three bills imposing term limitations on members of Congress.

April

5 Premier Jacques Parizeau of Quebec announces postponement of referendum on sovereignty until the autumn.

6 US House of Representatives approves a $180bn tax cut.

16 The EU and Canada reach agreement over fishing dispute. Spain to reduce its catch and accept stronger Canadian enforcement.

19   A 4,800lb car-bomb destroys a federal office building in Oklahoma City, killing 167.

May

2   US and Cuba sign immigration deal allowing 15,000 refugees from camps at the Guantánamo naval base to enter the US.
21   Former US Secretary of Defense Les Aspin dies from a stroke.

June

2   President Clinton renews China's most favoured nation status.
7–   House of Representatives rejects an attempt to repeal the 1973 War Powers Act; US Senate overwhelmingly approves anti-terrorist bill strengthening the government's power to prevent, investigate and punish terrorist acts at home and abroad (7); Clinton vetoes Congressional budget plan (7).

July

13   Clinton accepts proposal from the independent Base Closure and Realignment Commission to close 79 US military bases.
19   US House votes to restrict further disbursement of emergency loans to Mexico.

August

3   Iyad Mahmoud Ismail Najem, a suspect in the February 1993 World Trade Center bombing in New York, is extradited from Jordan to the US. Najem pleads not guilty to conspiracy charges.
15   US Air Force punishes seven officers involved in the friendly-fire downing of two US helicopters over Iraq in April 1994 in which 26 died.

September

7–   Parizeau proposes that referendum on Quebec sovereignty be held 30 October; date and wording of referendum approved by provincial legislature (20).

October

1   Ten convicted for the New York World Trade Center bombing by a federal jury.
9   Senator Sam Nunn, Chairman of the Armed Services Committee from 1987 to 1994, announces that he will not run for re-election in 1996.
19   Senate passes a bill tightening sanctions against Cuba.
30   Quebec votes against independence by less than 1%.

31   CIA Director John Deutch testifies before Congress that the Agency knowingly passed on suspect information regarding the Soviet Union to the US administration.

November

1   US Department of Defense announces the creation of a special panel to assess the impact of Soviet-distorted information on US military expenditures.
13–   President Clinton refuses to sign temporary funding provisions; nonessential government offices close (14); Congress and Clinton agree on temporary government funding, federal offices reopen (19).
27   Canadian Prime Minister Jean Chrétien announces proposals decentralising federal government functions, recognising Quebec as a distinct society and granting veto powers to four regions (Ontario, Quebec, the Atlantic provinces and the West) over any future Constitutional amendments.

December

16   A partial federal government shut-down begins as Congress and Clinton fail to reach a budget agreement.
28   Clinton vetoes a defence authorisation bill to develop anti-missile defence system.

Latin America

January

1   Mercosur, a free-trade area comprising Argentina, Brazil, Paraguay and Uruguay, is inaugurated.
1–   Fernando Henrique Cardoso takes office as President of Brazil; Cardoso's new government is inaugurated (2).
3–   Mexico's President Ernesto Zedillo Ponce de León announces economic programme, including wage increases, government spending cuts, privatisation and the lifting of exchange controls, to resolve the country's liquidity crisis; Minister of Interior Esteban Moctezuma Barragán and EZLN Sub-Commandant Marcos meet (15); ruling PRI and three opposition parties reach electoral agreement (17); IMF grants $7.75bn loan to Mexico (26); Clinton announces $50.76bn line of credit (31).
9   First reports of border clashes between Ecuador and Peru; Ecuadorian helicopters destroy Peruvian border post (26); Ecuadorian President Sixto Durán Ballén announces state of emergency (27); Peruvian forces attack four Ecuadorian military bases (29); Cease-fire announced, talks begin in Brazil (31).
12–   In Haiti, first US casualty as soldier is shot at a roadblock; UNSC passes Resolution 975 establishing UNMIH to take over military command from

US-led forces on 31 March.

24– Demobilised soldiers in El Salvador launch nationwide protest, seizing the Legislative Assembly, the Ministry of Finance, other government buildings and 2,500 hostages; hostages freed and government buildings handed over when government promises former troops land, credit, housing and job training (25).

February

2– Fighting erupts between Peru and Ecuador; Peru implements a unilateral cease-fire (14); Peru and Ecuador sign a peace declaration in Brasília (17); Fighting resumes near Tiwintza (22); cease-fire signed (28).

9– Mexican government ends truce with Zapatista rebels by sending troops into Chiapas to arrest rebel leaders; the government halts further military activities in the region (15).

20 Haitian President Jean-Bertrand Aristide dismisses all 43 military officers above the rank of major.

21 General Humberto Ortega retires as Commander-in-Chief of the Nicaraguan Army, replaced by his Chief-of-Staff, General Joaquin Cuadra Lacayo.

March

1 Julio María Sanguinetti inaugurated as President of Uruguay.

9 Zedillo announces austerity programme cutting government spending, raising taxes and energy prices, and limiting the increase in the minimum wage; US Congress passes bill setting out terms for negotiations with the rebel Zapatista movement (11).

11– Peru and Ecuador agree to conditions and schedule for the deployment of international observers to monitor the cease-fire; separation of forces begins (30).

15 Colombia and Venezuela deploy 11,000 troops near shared border.

31 US military command in Haiti handed over to UNMIH.

April

9 Alberto Fujimori re-elected President of Peru.

10 Haiti's Provisional Electoral Council postpones June legislative elections by three weeks.

18 Bolivian President Gonzalo Sánchez de Lozada announces 90-day state of emergency to quell civil unrest.

21–24 Zapatista rebels and Mexican government hold peace talks.

May

12–15 Second round of peace talks between Mexican government and Zapatista rebels held.

14 Carlos Menem re-elected President of Argentina.

30   Peru accepts border proposal mediated by Argentina, Brazil, Chile and US. Ecuador rejects plan.

June

7–11 Third round of peace talks held between Mexican government and Zapatista rebels.
9   Gilberto Rodríguez Orejuela, leader of the Cali drug cartel, arrested.
19–20 Presidential summit of Mercosur countries held in São Paulo, Brazil.
25–26 First round of local and legislative elections held in Haiti.

July

6–   Fourth round of talks between Mexican government and Zapatista rebels ends; Fifth round ends without any agreement (25).
8   Menem inaugurated as President of Argentina.
12–   Partial results of first round of Haitian elections shows Aristide's Lavalas movement winning all seats up for contention in the Senate and in the Chamber of Deputies. Opposition parties reject results in response to alleged widespread electoral irregularities; Provisional Electoral Council postpones the second round of elections from 23 July to 6 August (18).
18   Bolivia extends its state of emergency by 90 days.
25–   Peru and Ecuador sign an agreement demilitarising their border; Fujimori sworn in as President of Peru (28).

August

6   Miguel Rodríguez Orejuela, the Cali drug cartel's second-in-command, arrested.
13–   In Haiti, the first round of voting repeated in 21 districts; Provisional Electoral Council announces that the second round is to be held 17 September (23).
27   Zapatista rebels hold an unofficial plebiscite throughout Mexico to decide the movement's political future.

September

11   Mexican government and Zapatista rebels sign accord establishing working groups to discuss indigenous rights, social welfare, economic development and political reform.
27   Argentina and the UK sign agreement to begin joint oil and gas exploration in the South Atlantic near the Falkland Islands.

October

5–6   Peru and Ecuador hold talks in Brasília and agree to direct formal negotiations next year.

16   Haitian Prime Minister Smarck Michel resigns.
28   Colombian President Ernesto Samper Pizano denies any knowledge of proceeds from the Cali cartel to his 1994 campaign funds.

November

7   Claudette Werleigh sworn in as Haiti's Prime Minister.
12   Alvaro Arzú Irigoyen and Alfonso Portillo advance to the second round of Guatemala's presidential elections.

December

14   A Colombian congressional committee rules by 14 to 1 that sufficient evidence is lacking to begin an investigation of President Samper on charges that he accepted drug money for his 1994 campaign.
17   René Préval, a supporter of Jean-Bertrand Aristide, wins Haiti's presidential election.

Europe

January

1–   Four-month cease-fire in Bosnia begins; Bosnian Croats join cease-fire (2); renewed clashes in Velika Kladusa and Bihac (3); Croatian President Tudjman announces that UNPROFOR's mandate in Croatia will be terminated at the end of March (12); Bosnian government troops, backed by Croatian Army and Bosnian Croats, attack Bihac (14); General Rose replaced by Lt-Gen. Rupert Smith as UNPROFOR commander (26).
1–   Austria, Finland and Sweden join the EU; European Parliament approves new 20-member Commission with Jacques Santer as President (18); New Commission sworn in (24).
1–   Poland's currency, the *zloty*, is redenominated 10,000 to one; Foreign Minister Andrzej Olechowski resigns (13).
3–   Russia renews air and ground assault in Chechnya; Russian President Yeltsin orders cessation of all bombing on Grozny, but bombing continues (4); Russian forces attack presidential palace in Grozny (15); acting Prosecutor General Aleksei Ilyushenko announces criminal investigation of military leaders who allegedly refused to take command of Russian invasion (16); Presidential palace falls (19); Russian Human Rights Commissioner Sergei Kovaliev calls for an end to military operations in Chechnya (26).
12–   UK announces that the British Army will stop patrolling Belfast during daylight hours; UK Northern Ireland Secretary Sir Patrick Mayhew and Irish Prime Minister John Bruton meet in London to discuss all-party constitutional talks (26); Sinn Fein President Gerry Adams meets Bruton in Dublin (27).
25   Lamberto Dini endorsed as Italy's Prime Minister by the Chamber of Deputies.

February

1   Romania, Bulgaria, Slovakia and the Czech Republic gain associate EU status.

2–   Serbian President Milosevic and Bosnians reject French proposal for summit talks; Croats and Muslims agree to an international arbiter giving binding decisions on any dispute (5); Sarajevo airport road opened for first time in seven months to limited traffic (6); fighting intensifies in Bihac pocket (12); Serbian President Milosevic turns down Contact Group offer over sanctions (18); Bosnian and Croatian Serbs set up joint 'supreme defence council' (20); UN blockaded in camp at Visoko (24); aid convoy at Bihac abandoned under heavy artillery fire (28).

2–   Foreign Ministers of UK, France, Germany and Italy meet with Turkish Foreign Minister on EU entry and Cyprus; Greece drops veto on customs union with Turkey (6); Greek Cabinet rejects agreement over Turkish–EU customs union (9).

8   Russian–Ukrainian agreement over control and split of Black Sea fleet initialled by Presidents Yeltsin and Kuchma.

10   Austria joins as 25th member of NATO's Partnership for Peace.

12   Chechen General Mashkadov agrees to cease-fire suggested by Ingushetia to allow prisoner exchange; Russian forces renew offensive bombing attacks on Gudermes, Argun and Semashki; tanks move into Alkhan-Yurt, four miles south-west of Grozny (21).

13–   Chechen rebels and Russian forces agree to cease-fire; cease-fire takes affect (15); fighting resumes (19).

16   EU cancels meeting with 70 Asia Caribbean Pacific nations as no agreement on scale of aid package is reached.

16   Yeltsin confirms Russian parliamentary elections for December 1995 and presidential elections in June 1996.

March

1   Ukrainian Prime Minister Masol resigns, replaced by Evgeny Marchuk.

1   Russia repays $100m of $500m owed in interest arrears on commercial debt to London Club.

1   Polish parliament elects Jozef Oleksy to replace Pawlak as Prime Minister.

6–   Croatia forms military alliance with Bosnian Muslim-Croat federation; UN stops delivery of food supplies to Krajina (8); Tudjman agrees to UNPROFOR remaining in Croatia (12); British patrol comes under fire from Serbs near Gorazde (15); Muslim government troops break cease-fire and launch attacks on Tuzla and Travnik (20); weapons stolen by Serbs from UN collection site and from British troops (21).

14–   UK government reduces Northern Ireland garrison by one battalion; UK withdraws troops from Belfast patrols (26).

16   Slovak–Hungarian agreement reached on treaty on minority rights.

17   Ukrainian parliament votes to abolish Constitution and presidency of Crimea.

17   Azerbaijani troops crush mutiny by special police in Baku.

18   Finland returns Social Democratic Party to power.

20 Pan-European security conference, attended by 50 countries, opens in Paris.
21– Russia opens new offensive against Chechen rebel strongholds east of Grozny; Russian troops attack Shali, driving out most opposition troops (29); Russian Army takes Gudermes (30); Russian forces take Shali (31).
26 Fulfilling the Schengen Agreement, France, Germany, Benelux countries, Spain and Portugal lift controls on their common borders.
29 Germany suspends delivery of military equipment to Turkey; Turkey stops UN guards from patrolling.

April

1– Ukrainian President Kuchma announces decree putting Crimean government under his direct control and appointing Anatoli Franchuk Prime Minister; Ukrainian deputies vote 292 to 15 for no-confidence in the Cabinet (4).
4– Government troops take Mt Vlasik (Travnik); UN call in air cover over Sarajevo (10); Russian Duma votes overwhelmingly to lift UN sanctions on former Yugoslavia (14); Serbs use mortars against road over Mt Igman. NATO aircraft called in but no strike. Explosions in Serb suburb of Ilidza (Sarajevo) (17); UNSC, with Russia and China abstaining, resolves to maintain sanctions against Belgrade for 75 days (21); UN aircraft fired on while taking off from Sarajevo as cease-fire violations continue (24); Serb bombers based in Krajina attack target in Bihac enclave (29).
11 IMF approves $6.8bn loan for Russia.
16– Russian attack on Chechnyan town of Bamut repulsed; heavy fighting breaks out at Alleroi, close to Gudermes (23).
28 Austria joins Schengen group.

May

1– Croatian troops attack highway from the air. Serbs hold 120 UN officials and Croats surround Czech UN posts; Serb rocket attack on Zagreb kills five, injures over 120 (2); rocket attack on Zagreb (3); Bosnian Serbs shell the Sarajevan suburb of Butmir, killing 11 and injuring 14 (7); request for air-strikes after Serb mortar fire on Sarajevo vetoed by Special Representative Akashi (8); Serbs attack Orasje pockets (11); Croat tanks sent to aid Bosnian Croats under Serb attack at Orasje (14); shelling of Sarajevo continues and small artillery fired at Bihac (17); when Bosnian Serbs fail to meet UN ultimatum, ammunition dump at Pale attacked by six US aircraft, plus others – Serbs then shell Sarajevo, Bihac, Gorazde and Srebrenica, kill 70 in Tuzla (25); Serbs take about 300 UN troops as human shields (26); Lord Owen resigns as EU negotiator (31); Serbs attack Gorazde in attempt to move their lines closer to populated areas (31).
7– Jacques Chirac elected President of France; inaugurated (17); appoints new government with Alain Juppé as Prime Minister (17); Juppé announces new Cabinet (18).
10– UK government minister meets Sinn Fein delegation; Mayhew and Adams meet privately at US trade conference (24).

10   Italian troops take up position along eastern coastline to intercept Albanian refugees.

11–   In Chechnya, Russian Army shells village of Serzhen-Yurt and mountain strongholds; peace talks collapse after four hours (25).

23   Yeltsin vetoes Duma-passed law setting rules for the December parliamentary elections.

31   General Lebed, Commander of 14th Army in Moldova, resigns.

June

2–   US F-16 shot down by Bosnian Serb SAM; Serbs free 108 hostages, leaving some 150 still held (6); US House of Representatives votes 318 to 99 to end arms embargo on Bosnian government (8); US F-16 pilot recovered (8); Carl Bildt of Sweden replaces Owen as EU chief negotiator (9); Bosnian Croats mobilising near Livno 50 miles west of Sarajevo (15); Bosnian Serbs violate 'no-fly zone' twice, but UN refuses NATO authority to attack Banja Luka airfield (18); first aid convoys for over a week reach Sarajevo (18); Canadian UN troops blockaded by Bosnian government army for three hours in Visoko camp (20); elements of French and British Rapid Reaction Force (RRF) arrive in Split (22); Clinton diverts $50m from Department of Defense funds to RRF funding (29).

5–   Chechen rebels led by Shamil Basayev launch attack on Budennovsk, a Russian town near Chechen border (15); hold about 500 hostages in a hospital (16); Russian government loses vote of no confidence over handling of Budennovsk hostage situation (18); Chechen rebels allowed to leave Budennovsk with over 100 hostages including journalists and members of parliament (19); Rebels release all hostages at Zandak in Chechnya (20).

12   Estonia, Latvia, and Lithuania sign accords with EU.

29   Germany and Netherlands refuse French request for six-month delay in abandoning border controls.

30   Yeltsin accepts resignation of Ministers Yerin (Interior), Stepashin (Security Service), Yegorov (Deputy Prime Minister for Nationalities), but not that of Grachev (Defence).

July

2–   UN barracks in Sarajevo hit by three Serb shells and French respond to fire on Mt Igman; Srebrenica receives heaviest shelling since being declared a safe area (6); Serb tanks close on Srebrenica, over-run UN outposts and take 30 Dutch hostages (9); Zepa under artillery fire (11); Serbs take Srebrenica enclave, Dutch UN troops pull back to Potocari camp where refugees are congregating (11); Bosnian Serbs release about 130 UN hostages, holding only 15 (13); Serbs launch attack on Zepa (14); Serb forces on offensive in north-west Bosnia (20); Bosnian government rejects surrender terms, fighting resumes around Zepa after surrender deadline expires (20); Bosnian government and Croatia sign protocol on military cooperation (22); UK, US, French senior officers deliver ultimatum to General Mladic threatening robust response to attacks on all safe areas

(23); Foreign Ministers of eight Organisation of Islamic Conference states meet and agree to ignore arms embargo on Bosnia (23); Zepa captured. UN War Crimes Tribunal formally indicts Karadzic and Mladic with charges of genocide and crimes against humanity (25); Dole bill lifting arms embargo of Bosnia passed by Senate by 62 to 29 (26); Bosnian Croat forces claim to have captured 27 square miles south of Bihac in Livno region (27); Croatian forces cross into Bosnia, capturing Bosanko Grahovo and Glamoc. Bosnian Serbs halt attack on Bihac (28).

3   Rioting and fire-bombs in Belfast and Londonderry after Private Lee Clegg, who was serving a life sentence for shooting a teenage joyrider in September 1990, is released.

6   Yeltsin appoints Anatoly Kulikov, army commander in Chechnya, as Interior Minister.

11 Yeltsin admitted to hospital with suspected heart attack.

14 Yeltsin announces election date for lower house as 17 December 1995.

18 EU and Russia sign Trade Pact.

23 Turkish parliament approves constitutional reforms.

23– Russian and Chechen negotiators sign military agreement on prisoner exchange, communications, disarmament; Russia and Chechnya sign cease-fire (30).

August

1– US House of Representatives votes to lift Bosnian arms embargo; three Krajina Serb aircraft attack Croat troops at Strmica on border between Knin and Bosanko Grahavo (1); UN troops withdrawn from Zepa (2); Croatian forces shell Knin and other locations in Krajina (4); Karadzic dismisses Mladic as military commander but he refuses to step down (4); Bosnian Army makes gains in Bihac enclave, links up with Croats in Krajina (6); Bosnian government troops re-capture Muslim town of Velika Kladusa (7); Karadzic reinstates Mladic (11); Clinton vetoes bill lifting arms embargo (11); Milosevic replaces hardline Foreign Minister Vladislav Jovanovic with Milan Milutinovic (15); UN decides to withdraw troops from Gorazde leaving only a few observers (18); French 155mm guns on Mt Igman return fire on Serbs who wound six Egyptian soldiers in Sarajevo (22); UK troops attacked at Gorazde by Bosnian government troops (24); Serb mortar fire on Sarajevo kills 37 and wounds over 40 (28); Last UN troops in Gorazde leave and enter Serbia (29); NATO launches *Operation Deliberate Force*, attacking Serb positions with aircraft and artillery in response to shelling of Sarajevo (30); Bosnian Serb leader Karadzic agrees to join joint delegation with Serbia at peace talks (30).

2– First prisoner exchange between Russians and Chechens; first group of Chechens hands in arms (16); Chechens capture Argun police station, take hostages, and Russian Army storms building (21).

29 Georgian President Shevardnadze wounded by car-bomb in Tbilisi.

September

1– NATO aircraft resume bombing of Bosnian Serb positions; Bosnian Serbs respond by shelling Sarajevo (5); NATO aircraft attack Bosnian Serb targets near Pale (7); 13 *Tomahawk* cruise missiles fired at Bosnian Serb air-defence targets in the Banja Luka region from USS Normandy in the Adriatic (10); Bosnian government launches several attacks in central Bosnia (12); NATO aircraft continue to attack Serb targets near Sarajevo (13); Bosnian government troops in conjunction with Bosnian Croats and Croatian government troops open offensive on three fronts in central and western Bosnia (13); NATO suspends *Operation Deliberate Force* after Bosnian Serbs agree to withdraw heavy weapons from 20km exclusion zone around Sarajevo (14); UN relief flight lands in Sarajevo, first since April (15); Bosnian Serbs begin removing heavy weapons from around Sarajevo (16); Bosnian government forces, with Bosnian Croats and Croatian troops, capture large areas of northern Bosnia and reach within 30 miles of Banja Luka (18); UN breaks siege of Sarajevo using air and land routes to bring in supplies (18); UN/NATO indefinitely suspend threat to bomb Bosnian Serb military sites as heavy weapons withdrawn from the Sarajevo exclusion zone (21); attack on Serb-held Banja Luka halted (21); agreement signed in New York on framework constitutional arrangements for Bosnia-Herzegovina (26).

5– Irish government pulls out of scheduled 6 September summit with John Major; Gerry Adams meets US Vice-President Al Gore in Washington to seek US help in breaking the impasse over the future of Northern Ireland (14).

14 Greece and Former Yugoslav Republic of Macedonia sign agreement to establish diplomatic relations and economic ties.

14 Italy refuses US request to station F-117A *Stealth* aircraft at Italian airbase.

21 NATO ambassadors approve blueprint outlining membership requirements for former Communist Central and eastern European countries that want to join the alliance.

October

2– Serbs make battlefield gains in north-west Bosnia; Croats and Serbs strike deal over Eastern Slavonia to allow Croatian authority after a transition period (3); three Serb SAM sites attacked by NATO aircraft (4); cease-fire in Bosnia comes into effect (12); fighting continues in north-western Bosnia around Sanski Most and Prijedor and Serbs threaten to withdraw from cease-fire agreement (14); Bosnian Serb Prime Minister Dusan Kozic resigns and four Bosnian Serb generals dismissed following recent Serb losses (16); first civilian convoy in more than three and a half years drives through Serb-held Bosnia from Pazaric to Sarajevo (27).

4 Russian grain harvest worst in 30 years, down 19% from last year's total of 81.3 million tons.

6– Russia's military commander in Chechnya seriously injured in a bomb attack in Grozny; 18 Russian soldiers killed in an ambush on their convoy in Chechnya (26)

15   Turkish Prime Minister Ciller resigns after losing vote of no-confidence in parliament; calls for elections.
21   Willy Claes announces his resignation as NATO Secretary-General.
27   President Yeltsin admitted to hospital following second mild heart attack.
29   Franjo Tudjman's Christian Democratic Union wins over 45% of the vote in Croatia's parliamentary elections.
30   Chirac and Major complete two-day summit. Anglo-French air planning group promises cooperation on defence, naval and nuclear matters.

November

1–   Bosnian peace talks start in Dayton, Ohio; Croatia moves troops and heavy artillery towards border with Eastern Slavonia, the last Serb enclave on its territory (9); agreement signed in Dayton on re-integration of Eastern Slavonia into Croatia (12); peace agreement initialled by leaders of Serbia, Croatia and Bosnia (21); Security Council suspends sanctions on Yugoslavia (22); Bosnian Serb leadership meets Milosevic in Belgrade and accepts peace settlement (23).
5   Shevardnadze wins another five-year term in Georgian presidential elections.
7   Chirac dissolves government, Juppé resigns as Prime Minister and is immediately reinstated.
12   Reports that the UK might buy the US F-22 fighter aircraft raised concerns that the Eurofighter project could be affected.
19   Alexander Kwasniewski wins Polish presidential election.
27   President Yeltsin released from hospital and moved to a convalescent home.
28   UK and Irish Prime Ministers agree twin-track approach to peace in Northern Ireland.
28   European Union and 12 Mediterranean neighbours sign agreement in Barcelona on future political and economic cooperation.

December

1   Javier Solana Madariaga appointed NATO Secretary-General.
4–   Car-bomb in Grozny explodes outside the building of the Moscow-backed Chechen government, killing at least 11 and injuring 60; Chechen rebels launch fierce attack in Gudermes, heavy fighting lasts two weeks (14).
5   France announces decision to rejoin NATO's integrated military command.
5–   NATO formally endorses deployment of 60,000 troops to Bosnia; German Parliament votes to send 4,000 troops to Bosnia (6); two missing French *Mirage* aircrew released in good health by Bosnian Serbs after three months in captivity (12); in Paris, the Presidents of Bosnia-Herzegovina, Croatia and Serbia sign the General Framework Accord for peace in Bosnia-Herzegovina (14); US troops arrive in Tuzla (18); UN hands over command of military operations in Bosnia to NATO (20); rival armies pull back from cease-fire line in Sarajevo (22); US suspends economic sanctions

against Yugoslavia (28); NATO commander Admiral Leighton Smith turns down request by Bosnian Serb Assembly leader Momcilo Krajisnik for delay of at least nine months in transferring Bosnian Serb areas of Sarajevo to Muslim control (30).

17   Russia's Communist Party emerges as the biggest winner in elections to the Duma with 22.3 % of the vote and 157 seats in the 450-seat lower chamber, Prime Minister Chernomyrdin's 'Our Home' is Russia receives 10.1% and 55 seats, and Vladimir Zhirinovsky's far-right Liberal Democratic Party of Russia wins 11.2% of the vote, but only 51 seats.

20   UK and the US sign memorandum of understanding to start development of a new combat aircraft to replace the Harrier jump jet.

25   Islamic Warfare Party gains 21.3% of seats (158) in the Turkish parliamentary elections, Tansu Ciller's True Path Party gains 135, while the Motherland Party wins 132.

Middle East

January

1–   Israeli soldiers kill three Palestinian police officers in Gaza Strip; Israeli undercover unit shoot four members of the Popular Front for the Liberation of Palestine in Ramallah (4); PLO Chairman Yasser Arafat and Israeli Foreign Minister Shimon Peres meet at the Erez crossing point in the Gaza Strip, agree on recognition of Palestinian passports, creation of Palestinian licence plates, travel for students and men over the age of 50 between the Gaza Strip and Jericho, and cooperation in constructing a West Bank Industrial Park (9); Israel carries out air-strikes against *Hizbollah* bases in Southern Lebanon (10); Israeli Air Force strikes targets south of Beirut (15); two Israeli soldiers killed in southern Lebanon near village of Taibe (19); Islamic *Jihad* carries out a double suicide bomb attack in Netanya, killing 21 (22).

5–   Salam al-Majali resigns as Jordan's Prime Minister; Field Marshal Sharif Zaid ibn Shaker sworn in as replacement (8).

8   Iran signs deal with Russia to complete the construction of 1,300 megawatt nuclear power plant at Bushehr in southern Iran.

10–   PUK and KDP declare cease-fire in Iraq; clashes between PUK and KDP break out in Arbil, killing 100 (19).

15–   Yemen accepts Syrian mediation in border dispute with Saudi Arabia; talks begin in Riyadh (29).

30   UN renews sanctions against Iraq.

February

2–   Israel, Egypt, Jordan and PLO hold summit meeting in Cairo; Israeli Prime Minister Rabin and PLO Chairman Arafat meet at Gaza checkpoint (9); Israeli aircraft and artillery attack *Hizbollah* targets in south Lebanon

(18); Closure of Gaza-West Bank border eased allowing 1,000 into Israel (20); Israeli Air Force carries out two rocket attacks against *Hizbollah* targets (20).

27   Car-bomb kills 80 and injures more than 100 in KDP-controlled Iraqi town of Zakho.

March

6   France opens 'Interests Section' in Baghdad.
20–   35,000 Turkish troops cross Iraqi border in pursuit of PKK rebels; UN evacuates 2,500 refugees from Iraqi town of Zakho (25); Turkish forces drive Kurds towards Syrian and Iranian (Khakurk region) borders (27).
20   Qatar agrees to store US military equipment for one brigade.
26   Iraq rejects UN proposal for oil sales.
31   Israel attacks *Hizbollah* targets in south Lebanon.

April

2–   *Hamas* bomb factory explodes, killing one; two suicide bombers, one claimed by Islamic *Jihad*, the other by *Hamas*, kill seven, wound 46 Israelis in two car-bomb attacks in Gaza (9); Palestinian Security Court passes sentence on Islamic militant, and up to 150 militants arrested by PA following bomb attack (10); Israel bars Palestinian entry during Passover period (12).
8–   PUK and KDP agree on cease-fire; 20,000 (5 brigades) Turkish troops withdrawn from Northern Iraq (25).
30   Clinton announces cut-off of all trade and investment with Iran.

May

17–   US vetoes UNSC Resolution calling for Israel to reverse land expropriation; Israel suspends land expropriation (22).
19–   Rafiq al-Hariri resigns as Lebanon's Prime Minister; reappointed (21).
24   Syria and Israel agree on conditions for their Chiefs-of-Staff to meet under US auspices to discuss security arrangements in the Golan Heights.

June

5–   Israel arrests 45 *Hamas* members suspected of planning bombing campaign against Jerusalem; *Hizbollah* ambushes IDF in south Lebanon, killing three and wounding eight (18); leader of Islamic *Jihad*, Mahmoud al-Khawaja, assassinated in Gaza (22); Israel responds to *Hizbollah* rocket attack close to Nahariya with artillery and air attack (23); Israeli forces attack *Hamas* hide-out in Hebron (29).
14   Army mutiny against Saddam Hussein failed.
26–   Attempt to assassinate Egyptian President Mubarak fails; Egypt expels Sudanese from observation posts and formally accuses Sudan of Mubarak assassination attempt (29).
26   Emir of Qatar overthrown by eldest son; all GCC states recognise new

Qatari Emir (28).

26   Russia and Iraq sign deal for Russia to develop West Qurno and North Rumaila oilfields.

27–29   Israeli and Syrian chiefs of staff meet in Washington DC.

July

3–   *Hizbollah* kills two, wounds four Israeli soldiers with road-side bomb in south Lebanon; bomb in bus in Tel Aviv kills six, injures 30 (24); Knesset votes 59 to 59 on bill to prevent Israeli withdrawal from Golan (26).

4   Egyptian Islamic Group claims responsibility for Mubarak assassination attempt.

11   Turkish Army completes withdrawal from northern Iraq.

August

9–   Israel closes border with West Bank and Gaza for two days; *al Massar* ('the Path') group breaks away from *Hamas* and renounces violence (13); Palestinian shot by Israeli settlers as villagers protest at illegal settlement (13); bomb attack kills five and injures 107 in Jerusalem (21); protocol signed giving more powers to PA (27); Israel lifts closure of Jericho (30).

10   Two of Saddam Hussein's sons-in-law, with wives and families, defect to Jordan.

11   Israel and PLO initial Taba Declaration, providing for direct Palestinian elections and transfer of civilian authority from Israeli to Palestinian hands.

September

12   Libya begins to expel Palestinian workers and their families.

12   Senior Iraqi and Iranian officials hold talks in Tehran.

12–15   Inconclusive talks between the KDP and the PUK held in Dublin.

28   Israel and PLO sign Taba Agreement on Palestinian self-rule in the occupied West Bank.

October

1   Oman becomes first Gulf state to establish formal trade links with Israel.

10–   Israel releases 882 Palestinian prisoners; Israel performs search-and-destroy operation in south Lebanon following *Hizbollah* ambush on Israeli patrol which killed six (15); Israel and PLO agree on schedule for Israeli troop withdrawal (16); Israeli forces withdraw from West Bank towns starting with Jenin (26).

12–   Three Israeli soldiers killed by land-mine in southern Lebanese town of Aishiyeh; six Israeli soldiers killed by roadside mine near village of Rayhan in South Lebanon.

15   Saddam Hussein receives 99.96% of vote in national referendum.

23–   Ten killed and at least 80 injured in two car-bomb attacks in Algiers; six dead and 83 wounded in bomb blast in Rouiba (29).

26   Founder and leader of Islamic *Jihad* assassinated in Malta.

November

2–   Two suicide bombers killed and 11 Israelis injured in two bombings in the
       Gaza; Israeli Prime Minister Rabin shot dead by a Jewish right-wing
       extremist who is immediately arrested (4); Israeli soldiers withdraw from
       West Bank town of Jenin (13); new Prime Minister Peres announces
       Cabinet in which he has the Defence portfolio and Ehud Barak is Foreign
       Minister, and Haim Ramon is Interior Minister (21); *Hizbollah* guerrillas
       fire rockets into northern Israel (28).
13   Bomb attack near Saudi Arabian National Guard base in Riyadh kills
       seven people.
16–   Liamine Zeroual wins Algeria's first multi-party presidential elections;
       Zeroual sworn in (27).
20–   Car-bomb outside Egyptian Embassy in Islamabad kills 17; Egyptian
       government jails 54 members of Muslim Brotherhood and shuts Cairo
       headquarters (23); in the first round of parliamentary elections, Egypt's
       ruling party, NDP, wins 123 seats in the 444-seat National Assembly (29).

December

4   King Fahd of Saudi Arabia suffers stroke.
6   In second round of Egyptian elections, ruling NDP wins 416 out of 444
       seats.
17–   Eritrea and Yemen battle over disputed Greater and Lesser Hanish islands
       in the Red Sea; agreement reached on the release of 195 Yemenis taken
       prisoner by Eritrean forces in the fighting (27).
22   US Congress passes bill imposing penalties on foreign companies which
       invest in Iran's oil industries.

Asia and Australasia

January

1–   North Korea lifts ban on import of US products and arrival of US mer-
       chant ships; US partially lifts economic sanctions against North Korea (20).
12   Prototype of US/Japanese FSX support fighter unveiled.
13   Taiwanese cabinet approves plan to ease prohibition on direct links with
       mainland China.
18–   US–China talks on copyright infringement resume; talks end without
       agreement (28).
22   Myanmar troops capture Karen National Union stronghold at Manerplaw
       on Thai border.
28   US and Vietnam sign agreement for liaison offices in each others capitals.

February

2   South Korea rejects offer of political party talks with North Korea before government-to-government talks.

5   China retaliates against US trade sanctions by imposing tariffs on US goods including cigarettes and alcohol.

8   US liaison office in Hanoi opens.

8–   Philippines make formal protest after Filipino fisherman detained and Chinese military deployment on Mischief Reef violates 1992 Manila Declaration; Philippine government orders military to reinforce naval presence at Panganiban reef (15).

16– Thai forces put on alert after Myanmar incursions in hot pursuit of Karen rebels. Thai Army takes Karen base at Kawmoora (21).

26   China and the US sign agreement on protection of intellectual property.

28   Polish armistice delegation forced to leave North Korea.

March

2   India and China begin talks to settle Himalayan border dispute.

9–   Accord for establishing Korean Peninsula Energy Development Organisation signed in New York although North Korea still refuses South Korean reactors; North Korean proposal at Berlin talks that US reactors be provided rejected by South Korea (28).

17   One-third of deputies at CCP national congress unexpectedly withhold support for Jiang Chunyun, government nominee as vice-premier – Jiang Zemin nevertheless emerges with new leadership structure essentially intact.

20   Sarin nerve gas attack on Tokyo underground kills 7, injures 4,700.

20– China and Philippines begin talks on disputed islands; Philippine Navy destroys Chinese marker buoys and seizes four Chinese fishing boats near Spratly Islands (25).

21   Members of rebel group, Karen National Progressive Party, hand in over 8,000 weapons.

31   Taiwan sends three patrol boats to the Pratas and Spratly Islands to protect fishing fleet.

April

4   Several hundred Philippine Muslim rebels attack southern town of Ipil.

12   Chinese Prime Minister Li Peng announces that Jiang Zemin is confirmed as Deng Xiaoping's successor.

19– 370 in hospital after gas attack at Yokohama train station in Japan; Second gas attack at Yokohama train station hospitalises 25 (21).

26   Mahathir Mohamed's National Front wins Malaysian parliamentary election with large majority.

May

3   North Korea bars Neutral Nations Supervisory Committee staff from

entry to Panmunjon.

5   Thai forces attack Karen camp in Myanmar.

8   Pro-Ramos candidates win Philippine Congressional elections.

13   Confrontation between Chinese fishing boats and navy, and Philippines naval ship at Mischief Reef.

15   Shoko Asahara, leader of *Aum Shinrikyo* cult, arrested for sarin gas attacks.

June

13– Taiwan's President, Lee Teng-hui, visits US; Chinese Ambassador to Washington recalled. Beijing calls off July talks with Taiwan (17).

18– North Korea agrees to accept 150,000 tons of South Korean rice; South Korea suspends rice delivery as ship in port made to fly North Korean flag (29).

20   Talks begin between Philippine government and Muslim separatists who demand control of southern islands.

22   Japan formally decides to provide rice for North Korea.

22   US gives assurance to Japan that it will provide naval protection if South China Sea trade routes are threatened.

July

4   Two cyanide bombs found and defused in Tokyo subway stations.

5   Cambodian government admits that Khmer Rouge captured Treng (40km south of Battambang).

10   In Myanmar, Aung San Suu Kyi released from house arrest.

13   US reaffirms its policy of recognising one China if achieved through peaceful agreement between China and Taiwan.

August

4   ASEAN announces agreement with China to exchange defence information.

6   US Embassy opens in Hanoi.

10– China and Philippines agree 'code of conduct' over Spratly Islands dispute to avoid use of force.

23– In China, Harry Wu is sentenced to 15 years imprisonment for spying; he is deported immediately after sentencing (24).

24   Japanese fighters scramble as Chinese aircraft nears Japanese airspace close to disputed Senkaku Islands (100 miles north-east of Taiwan).

September

4– Three US servicemen allegedly rape 12-year-old Japanese girl in Okinawa; arrested and indicted (29).

5   North and South Korea agree to resume the talks suspended over rice aid spy ship allegation.

18   Hong Kong elections victory for Martin Lee's Democratic Party and Beijing rejects the poll.

20   UN rejects Taiwan's application for membership.
22   Ryutaro Hashimoto elected president of Japan's LDP.
25–28 Plenary session of the CCP's Central Committee approves China's ninth five-year plan.

October

3   Japan signs deal to send 200,000 metric tons of rice to North Korea.
5   General Fu Quanyou replaces General Zhang Wannian as China's Chief of the Army Staff.
11–   Japanese Premier calls for US troop reductions in forces based in Okinawa; 85,000 Okinawans protest against US military bases on island (22).
13   President Jiang Zemin arrives in South Korea for a five-day visit, the first by a Chinese head of state.
16   Taiwan gives cautious but positive response to reported offer by Chinese President Jiang Zemin for an exchange of visits by the presidents of the two countries.
19   At the APEC Conference in Osaka, China announces trade reforms, including reduced tariffs and import controls.
26   Taiwan purchases *Mistral* SAM from France.
29   Peace talks between the Indonesian government and Muslim rebels begin in Jakarta.

December

1   South Korea accuses North Korea of massing war planes near border.
3   South Korea's Chun Doo-hwan arrested on charges of staging an army mutiny in 1979.
3   In Taiwan's legislative elections, governing Nationalist Party wins narrow majority as Chinese New Party makes largest gains.
13   Wei Jingsheng jailed by the Chinese for an alleged plot to topple the government.
15   The leaders of all ten South-east Asian countries (including Burma, Cambodia and Laos) meet for the first time in Bangkok.
18   Indonesia and Australia agree on closer military links and mutual assistance.
27   China creates 150-member Preparatory Committee, which includes 94 Hong Kong and 56 Chinese officials, but excludes Hong Kong Democratic Party representatives, for Hong Kong July 1997 turnover.

South and Central Asia

January

2   Nine Russian border guards killed in eastern Gorny Badakhstan in Tajikistan in clash with rebels based in Afghanistan.

3– Peace talks resume between Sri Lankan government and the LTTE in Jaffna; two-week cease-fire begins (8); new round of talks begin in Jaffna (14).

10 Pakistani Prime Minister Benazir Bhutto calls on US to deliver the 32 F-16s Pakistan has already paid for or refund $650m.

12– Draft agreement on closer military cooperation between the US and India signed in New Delhi; India rejects Pakistani pre-condition of significant troop reduction in Kashmir for bilateral talks (16).

20– Russia and Kazakhstan sign military agreement to unite armed forces by the end of the year; agree to cooperate in oil and gas exploration in the Caspian Sea (23).

22 President Islam Karimov's People's Democratic Party and its allies wins control of Uzbekistan's Supreme Assembly in legislative elections.

23 Week-long cease-fire in Afghanistan breaks down.

February

5– First round of elections for Kyrgyzstan's new bi-cameral legislature; second round takes place (19).

10 Presidents of Kazakhstan, Kyrgyzstan and Uzbekistan establish Inter-state Council to govern 1994 trilateral economic union.

14 In Afghanistan, *Taleban* forces capture stronghold of ex-Prime Minister Hekmatyar's *Hezb-i-Islami* in Charasiab.

26 First round of legislative elections in Tajikistan are boycotted by the opposition and condemned by the UN and the OSCE as undemocratic.

March

6– Government troops launch heavy attack in south-west Kabul against Hekmatyar forces; *Taleban* makes first attack on Kabul, after taking over some of Hekmatyr positions south of the city (8); government troops force both *Taleban* and *Herb-i-Wahdat* out of Kabul (12); government troops capture *Taleban* base at Charasyab 15 miles south of Kabul (19); *Taleban* launches successful counter-attack against government forces south of Kabul (21).

6– Kazakhstan's Constitutional Court annuls the country's 1994 legislative elections on procedural grounds; President Nursultan Nazarbayev dissolves parliament (11).

26 In Uzbekistan, referendum backs the extension of Islam Karimov's presidential term by three years – next election to be held in 2000.

April

5–16 Pakistani President Benazir Bhutto carries out successful visit to the US.

7 Fighting breaks out between Tajik rebels and government troops supported by Russian border guards.

17–19 Iranian President Ali Akbar Rafsanjani visits India.

19– In Sri Lanka Tamil Tigers attack two army posts; police arrest 1,000

Tamils across country (23); Tamils shoot down two government trans-
ports (29).

30   In Kazakhstan, President Nazarbayev wins 95% of the vote to extend his
      term to 2000.

May

5   Nepal's army commander, General Gadul Shamsher Rana, resigns, re-
      placed by General Dharma Palbar Singh.
8–   Tamil Tigers kill 19 commandos near eastern town of Amparai; Air Force
      attacks Tamil Tiger positions south of Elephant Pass and Pooneryn army
      base (10); Tamil Tigers attack army post, leaving more than 60 dead (28).
10   Tajikistan introduces its own rouble currency.
11   Indian Army destroys Islamic shrine in Kashmir, killing 30 armed separatists.
18–   Violent clashes between the police and the pro-Urdu *Mohajir Qaumi Move-
      ment* (MQM) break out in Karachi; at least 20 die in a general strike called
      by the MQM (22); 13 more killed in various parts of the city (29-30).

June

9   Afghan government signs 10-day truce with *Taleban* rebels.
13   Nepal's King Birenda dissolves parliament.
24–   Twenty-eight killed in clashes with security forces during MQM strike in
      Karachi; President Bhutto announces formation of a committee to hold
      talks with MQM (27).
28   Tamil Tigers raid Mandaithivu Island.

July

4   Nazarbayev publishes new Kazakh draft Constitution.
9   Sri Lankan Army launches major offensive against Tamil Tigers.
11–   Talks begin between Pakistani government and MQM; negotiations col-
      lapse (31).
21   The UN peace envoy, Mahmoud Mestiri, visits President Burhanuddin
      Rabbani of Afghanistan in Kabul.

August

3–   Twenty-seven killed in violence in Karachi; 14 killed in further clashes (15);
      two-day general strike in Karachi leaves eight dead (23-24); talks resume
      between MQM and Pakistan government (29); negotiations adjourned (30).
4–   Sri Lankan government announces constitutional reforms giving Tamils
      greater autonomy in north and east; Tamil Tigers hijack ferry and sink
      two naval patrol boats sent to investigate (30).
20   China and India reach agreement to withdraw from four border posts in
      the Sumdarung Chu valley.
31   President Nazarbayev wins 89% of votes in referendum to give him more
      power.

September

3–   Talks with MQM resume; two MQM members shot dead (25); MQM breaks off talks (27).
4   Pro-Pakistani rebels explode bomb in Kashmir which kills 15.
5   In Afghanistan, *Taleban* captures city of Herat from forces loyal to President Rabbani.
10–   LTTE kill seven soldiers in the eastern district of Batticaloa; Sri Lankan government launches new offensive against the Tamil Tigers (12).
15   Nazarbayev issues decree moving Kazakhstan's capital from Almaty to Akmola in the north.

October

11–   *Taleban* seize the town of Charasyiab, near Kabul; capture Baiman, but lose the Sanglakh valley to government forces (15); launch rocket attack on Kabul market, killing 11 (21).
14–   Pakistan government announces arrest of army officers who planned to stage a coup; MQM stage strike (12); 11 killed in violence in Karachi (30).
19–   In Colombo, Tamil Tigers bomb Sri Lanka's only oil refinery; Tamil Tigers destroy oil storage tanks at two main depots holding all petrol brought into the country, and massacre 66 in attacks on three villages in the northeast (21); Sri Lankan government forces capture LTTE town of Neerveli (29); More than a 100,000 flee Jaffna as Sri Lankan Army continues its offensive (31).

November

10–   Government forces resume offensive against Tamil Tigers in Jaffna; seize rebels' political headquarters in Kodavil (12); capture LTTE training camp (14).
21   More than 20 injured in New Delhi bomb blast.
24   Bangladesh President Biswas dissolves parliament, asks Prime Minister Khaleda Zia to continue as head of an interim government.
26   *Taleban* air-strikes on Kabul kill at least 35.

December

5–   Elections to Kazakhstan's Senate; first round of elections to the *Majlis*, the Lower House, takes place (9); second round of elections held (23).
6–   Sri Lankan troops capture Jaffna; Government launches offensive against the Tamil Tigers in the east (10); LTTE ambush army patrol in the Batticaloa region, killing 33 government soldiers and losing 60 of its own (23).
12–   *Taleban* forces bombard Kabul with artillery; Fighting escalates to the north and south of the capital (19).
24   Kyrgyzstan's President, Askar Akayev, re-elected.

Africa

January

6   Joe Slovo, Chairman of the South African Communist Party and Minister of Housing, dies of cancer.

7–   Truce between Niger's government and Tuareg rebels extended by three months; opposition parties to the ruling Alliance of the Forces for Change win 43 out of 83 seats to the National Assembly (12).

10   Angolan Chief of Staff, General Arlindo Chenda Pena Ben-Ben, meets his UNITA counterpart, João Matos, and agree on cessation of hostilities, prisoner of war release, and separation of troops.

13–   UNSC extends UNOMIL's mandate to mid-April; Liberian peace negotiations adjourn indefinitely (31).

24–   UN abandons plan for multinational peacekeeping force to protect Rwandan refugees in Zaire; UNHCR agrees to pay $13.7m for 1,500 Zairian troops to protect refugees (29).

25   In Lesotho, Moshoeshoe II restored to throne after voluntary abdication by his son, Letsie III.

February

2   Uganda deploys troops on Sudanese border after Sudanese air attack on Kitguz.

8–   UNSC approves new peacekeeping mission to Angola of up to 7,000 troops but only 2,800 to be sent immediately; UNITA National Congress votes to keep peace accord (12).

26–   Inter-clan fighting increases around Mogadishu's airport in anticipation of UN's withdrawal from Somalia; US and Italian Marines land to cover UN withdrawal from Somalia (27); UN troops start withdrawal from airport while Somali militia fails to stop looting (28).

March

2   UN withdrawal from Somalia completed without casualties or clashes.

5   Zulu *Inkatha* Freedom Party ends boycott of South African parliament.

18   Algerian troops launch offensive against the GIA.

19   Sierra Leone Army, backed by Ukrainian helicopter pilots, re-takes southern town of Moyamba from Revolutionary United Front.

20–   Seventeen killed including three Belgians in clashes between Tutsi and Hutu in Burundi; Major increase in movement of Hutu refugees reported as 40,000 move to Tanzania (30).

27   President Mandela dismisses wife Winnie Mandela (Deputy Minister of Arts, Culture, Science and Technology) from cabinet.

31   Tanzania closes border to Burundi refugees.

April

8   *Inkatha* Freedom Party pulls out of South Africa's Constitutional Assembly.

8   Koibla Djimasta appointed Prime Minister of Chad.

8–9   Legislative elections held in Zimbabwe, and the ruling Zimbabwe African National Union-Patriotic Front wins 118 out of 120 seats to the National Assembly.

12   Gabriel Koyambounou appointed Prime Minister of the Central African Republic.

13   UNSC extends UNOMIL's mandate until 30 June.

17–   Rwandan government blocks aid convoys to refugee camps in Zaire (Kigali) and forces refugees out of Kibeho camp in South-west Rwanda; 2,000 massacred in Kibeho camp (22).

May

1   In Sierra Leone, Freetown virtually surrounded by Revolutionary United Front rebels.

5   Ethiopian People's Revolutionary Democratic Front wins overwhelming majority in the county's first multi-party legislative elections.

7   UNITA leader Savimbi and Angolan President dos Santos meet in Lusaka, Zambia.

11–   WHO confirms that Ebola virus has broken out in and around the Zairian town of Kikwit, population of 1,000,000; Kikwit quarantined (17).

June

2   In Sudan, SPLA accepts two-month extension of cease-fire.

6–   Burundi government troops, mainly Tutsi, surround suburb of Kamenge in capital, Bujumbura, where Hutu militia prepares to defend enclave; government troops take control of Kamenge without fighting (7).

11   Guinea's ruling Party of Unity and Progress wins the country's first multi-party legislative elections.

27   Nigeria's General Abacha lifts ban on political parties.

July

11   Sheikh Abdel Baki Sahraoui, founder of Islamic Salvation Group, shot dead in Paris.

16   Boutros Boutros-Ghali visits Angola. Savimbi holds press conference with UN Secretary-General.

29–30 ECOWAS meeting held in Accra, Ghana.

August

13   Burkina Faso's government announces capture of town of Kaya in West Equatoria province.

15   Army officers stage bloodless coup in São Tome and Principe.
16–   UN lifts arms embargo on Rwanda for one year; Zairian Army starts to
      send refugees back to Rwanda from Goma and Muganga (21); UNSC
      instructs Zaire to stop forced repatriation of refugees (22); Zaire agrees
      (24); UN fails to persuade refugees to return voluntarily from Zaire.
20   In Nigeria, main three factions sign peace agreement.
26–29 At least 22 killed and 150 injured in inter-clan fighting in Mogadishu,
      Somalia.

September

11   Rwandan Army kills over 100 in revenge attacks in Kanama, near the
      Zairean border.
12–14 At least two killed and 30 wounded in anti-government riots in Khar-
      toum, Sudan.
17   In Somalia, Mohammed Aideed captures town of Baidoa, threatening
      renewed civil war.
22   New Ugandan Constitution enacted.
29   EU–Moroccan fish talks collapse.

October

1   Nigeria's ruler General Abacha announces extension of military rule for a
      further three years.
1–   Algerian militants kill 18 in an attack on a bus carrying civilians; Algerian
      government forces kill 100 rebels in operations in the mountains east of
      Algiers (1–11).
3–   French land 900 troops on Comoros Islands; Bob Denard's coup attempt
      fails (4).
23   Meeting in New York between Presidents Chirac of France and Zeroual of
      Algeria called off at the last moment.
31   In Nigeria, Ken Saro-Wiwa and eight others sentenced to death by a
      military court.

November

1   First all-race local elections take place in South Africa.
4   Rwandan government soldiers kill 300 former Rwandan Hutu troops and
      militiamen in an assault on Iwana island in Lake Kivu near Zairean
      border.
10–   The Nigerian government executes dissidents; Nigeria suspended from
      the Commonwealth and sanctions proposed (11); UK and Netherlands
      reject proposal for Nigerian oil embargo at EU Foreign Ministers' meet-
      ing, while enforcing arms embargo and other sanctions (20); US decides
      against US oil embargo (22).
18   In Algerian election, President Zeroual wins 61% of vote in a 75% turnout.
26   Libyan opponent of Gaddafi regime, Ali Mehmed Abuzed, murdered in
      London.

December

1   In South Africa, General Malan and 19 others formally charged with the murder of 13 people in 1987.
6–   Rwandan government expels 43 NGOs from the country for failing to register; eight of the 43 organisations informed that they will be allowed to stay in Rwanda (15).
11   UK expels Libyan diplomat and Libya reciprocates.
12   UNSC votes unanimously to reduce its peacekeeping force in Rwanda by one-third and to end the two-year mission in March 1996; Rwanda war crimes tribunal issues first international arrest warrants for eight men accused of genocide and crimes against humanity.
12   Fifteen people killed and more than 30 injured when a car-bomb explodes in an Algiers suburb.
17   Cape Verde's ruling Movement for Democracy wins large majority in legislative elections.
21   Opposition alliance wins 60 out 66 seats in elections to the Mauritian National Assembly.

International Organisations/Arms Control

January

1   WTO, successor to GATT, comes into operation.
9–   Talks convene between US, South Korea, North Korea and Japan on the creation of Korean Energy Development Organisation; first shipment of US oil to North Korea arrives (17); second shipment arrives (19).
13   1993 UN Chemical Weapons Convention fails to be enacted since only 20 countries ratify the treaty (65 are needed).
26   UN commission recommends sweeping reform of UN's organisation, including creation of an 'economic security council' and the dismantling of some agencies.
31   US announces one-year extension of its unilateral moratorium on nuclear testing.

February

4–5   Finance Ministers and central bank governors of the Group of Seven countries meet in Toronto.
8   NATO approves plans for direct dialogue with Egypt, Israel, Mauritania, Morocco and Tunisia to combat the threat of Islamic fundamentalism.
19–   Rolf Ekeus, Head of UNSCOM, arrives in Baghdad; Ekeus ends current mission (23).
23   Belarus suspends its weapons destruction programme, violating the CFE Treaty.

March

2   Clinton announces removal of 200 tons of fissile material from stockpile.
2–   Belgium bans production and use of anti-personnel mines; UK government bans export of non-detectable anti-personnel mines (15).
13   UNSC renews sanctions on Iraq as all resolutions not complied with.
26   North Korea and US complete two days of talks in Berlin.
30   UNSC renews sanctions against Libya over Lockerbie bombing suspects.

April

6–   UK drops its demand for CTBT to allow special safety tests; UK announces its commitment to end production of fissile material for weapons (18).
11   UNSC adopts Resolution 984 guaranteeing support for non-nuclear states under nuclear threat or attack.
17   China rejects US request not to sell Iran nuclear reactors.
20   North Korean/US nuclear talks end without agreement.

May

10   US, South Korean and Japanese officials meet in Seoul.
11   NPT extended indefinitely by consensus.
11   North Korea agrees to resume talks and maintain nuclear freeze until end of talks.
28   China suspends talks with US on missile technology controls and cooperation on nuclear energy, postpones visits by US officials over Taiwanese President's visit to US.

June

1   Greece ratifies UNCLOS.
12   IAEA governors approve measures to increase effectiveness of inspectors.
12   Negotiators reach agreement over North Korean nuclear programme, with North Korean acceptance of light-water reactors from South Korea.
15   Australia and New Zealand freeze military cooperation with France over its decision to resume Pacific nuclear tests.
21   START II Treaty submitted to Russian parliament for ratification.
22   CIA report states China has recently delivered missile components to Iran and Pakistan.

July

5   UN conference on mine clearance begins in Geneva.
6   Belarus President Lukashenka orders a halt to the return of SS-25s to Russia (18 remain).
9   French commandos board *Rainbow Warrior* in Mururoa and arrest Greenpeace members.

20   Japan formally protests against French plans for nuclear testing in the Pacific.

August

17–   China carries out nuclear test; Japan freezes aid to China over its nuclear test (30).
28   Greenpeace helicopter violates 12-mile zone as protest fleet assembles near Mururoa.

September

2   French commandos board and tow away two Greenpeace ships.
5–   France detonates 20kt nuclear weapon under Mururoa atoll; Chirac postpones state visit to Tokyo and Swedish Prime Minister Ingvar Carlsson's visit to Paris (7); Severe riots in Tahiti over French testing (7);
12   Framework for the Wassenaar Arrangement, the successor to COCOM, establishing export controls on certain conventional weapons and dual-use technologies such as advanced computers and telecommunications equipment, reached in The Hague.
17   US indicates will soon sign NWFZ treaty in the South Pacific after years of opposition.
21   China blocks Taiwanese membership of the UN.
21   US Senate approves delivery of $368m-worth of arms to Pakistan.
25   Review Conference of UN Convention on Conventional Weapons opens in Vienna.

October

2–   France detonates second nuclear device at Mururoa atoll; European Commission rules that there are no grounds for action against France under the EU Atomic Energy Treaty (25); France carries out third nuclear test (27).
13   UN CCW Conference ends without a global ban on anti-personnel landmines.
18–20 Summit meeting of the NAM held in Cartagena de Indias, Colombia.
22   World Leaders meet in New York to celebrate UN 50th anniversary.

November

10–   UK criticised at the Commonwealth conference in Auckland for supporting France's nuclear testing programme; First Committee of the UN General Assembly passes a resolution condemning nuclear testing (16); Chirac cancels summit meeting with Italian prime minister and postpones meeting with Belgian prime minister (18); France conducts fourth nuclear test at Mururoa atoll (21).
13   The US Senate Foreign Relations and Armed Services Committees blocks ratification of the Chemical Weapons Convention.

14   The US agrees to allow Russia to expand its defences against nuclear missiles without violating a key arms-control agreement.

17   Russia fails to comply with the terms of the CFE Treaty.

December

15   North Korea signs deal with the Korean Peninsula Energy Development Organisation on the supply of nuclear reactors.

15   ASEAN countries sign treaty establishing a nuclear-weapon-free zone in South-east Asia.

19   Twenty-eight countries sign the Wassenaar Arrangement.

21   Czech Republic becomes first post-communist state to join OECD.

27   France conducts fifth nuclear test of current series under the Mururoa atoll.

Glossary

ABM	Anti-Ballistic Missile
AIS	Islamic Salvation Army (Algeria)
ANC	African National Congress
APEC	Asia-Pacific Economic Cooperation
ARF	ASEAN Regional Forum
ASEAN	Association of South-east Asian Nations
AWACS	Airborne Warning and Control System
BJP	*Bharatiya Janata* Party (India)
BTWC	Biological and Toxic Weapons Convention
BW	Biological Weapons
CCP	Chinese Communist Party
CCW	Convention on Conventional Weapons
CD	Conference on Disarmament
CFA	Court of Final Appeal (Hong Kong)
CFE	Conventional Forces in Europe
CFSP	Common Foreign and Security Policy (Europe)
CIS	Commonwealth of Independent States
CJTF	Combined Joint Task Force
COCOM	Coordinating Committee for Multilateral Export Controls
CSU	Christian Social Union (Germany)
CTBT	Comprehensive Test Ban Treaty
CUF	Civic United Front (Tanzania)
CW	Chemical Weapons
DPP	Democratic Progressive Party (Taiwan)
ECMM	European Union Monitors
ECOMOG	ECOWAS Peacekeeping Force in Liberia
ECOWAS	Economic Community of West African States
ECR	Electronic Combat and Reconnaissance
ECU	European Currency Unit
EMU	Economic and Monetary Union
EO	Executive Outcomes mercenary group (Africa)
EPA	Economic Planning Agency (Japan)
ESDI	European Security and Defence Identity
EU	European Union
EZLN	*Ejército Zapatista de Liberación Nacional* (Mexico)
FDP	Free Democratic Party (Germany)
FIS	Islamic Salvation Front (Algeria)

FLN	National Liberation Front (Algeria)
FRY	Federal Republic of Yugoslavia
FYDP	Future Years Defense Program (US)
FYROM	Former Yugoslav Republic of Macedonia
GATT	General Agreement on Tariffs and Trade
GCC	Gulf Cooperation Council
GDP	gross domestic product
GIA	Armed Islamic Group (Algeria)
GNP	gross national product
GPS	Global Positioning System
HRAT	UN Human Rights Action Team
HVO	Bosnian Croat Army
IAEA	International Atomic Energy Agency
ICBM	Inter-Continental Ballistic Missile
ICRC	International Committee of the Red Cross
ICTY	International Criminal Tribunal for the Former Yugoslavia
IDF	Israeli Defence Force
IFOR	NATO Implementaion Force (Bosnia)
IGC	Inter-Governmental Conference (Europe)
IMF	International Monetary Fund
INC	Iraqi National Congress
INF	Intermediate-range Nuclear Forces
IRA	Irish Republican Army
JAST	Joint Advanced Strike Technology (US)
KDP	Kurdish Democratic Party (Iraq)
KMT	Kuomintang (Taiwan)
LDP	Liberal Democratic Party (Japan)
LEGCO	Hong Kong Legislative Council
LTTE	Liberation Tigers of Tamil Eelam (Sri Lanka)
MINUGA	UN Verification Mission in Guatemala
MPLA	*Movimento Popular para a Libertação de Angola*
MQM	Mohajir Qaumi Movement
MRAV	Multi-Role Armoured Vehicle
NAC	North Atlantic Council
NAFTA	North American Free Trade Agreement
NAM	Non-Aligned Movement
NATO	North Atlantic Treaty Organisation
NDP	National Democratic Party (Egypt)
NFP	New Frontier Party (Japan)
NGO	Non-Governmental Organisation
NPT	Nuclear Non-Proliferation Treaty

NPTREC	NPT Review and Exyension Conference
NWFZ	Nuclear-Weapon-Free Zone
OAS	Organisation of American States
OAU	Organisation of African Unity
OECD	Organisation for Economic Cooperation and Development
OPEC	Organisation of Petroleum Exporting Countries
OSCE	Organisation for Security and Cooperation in Europe
PA	Palestinian Authority
PAN	National Action Party (Mexico)
PDD	Presidential Decision Directive (US)
PFP	Partnership for Peace (NATO)
PKK	Kurdish Workers' Party
PLA	Peoples' Liberation Army (China)
PLO	Palestine Liberation Organisation
PNC	Palestinian National Council
PRD	Democratic Revolution Party (Mexico)
PRI	Institutional Revolution Party (Mexico)
PUK	Patriotic Union of Kurdistan (Iraq)
R&D	Research and Development
RMA	Revolution in Military Affairs
RPF	Rwandan Patriotic Front
RSK	Serb Republic of Krajina
SAARC	South Asian Association of Regional Cooperation
SAM	Surface-to-Air Missile
SDF	Self-Defense Forces (Japan)
SDPJ	Social Democratic Party of Japan
SEANWFZ	South-East Asia Nuclear-Weapon-Free Zone
SHAPE	Supreme Headquarters Allied Powers Europe
SIVAM	Amazon Monitoring System
SPD	Social Democratic Party (Germany)
SPLA	Sudan Peoples' Liberation Army
START	Strategic Arms Reduction Talks
tesobonos	public debt bonds (Mexico)
THAAD	Theater High Altitude Area Defense (US)
UN	United Nations
UNCLOS	UN Convention on the Law of the Sea
UNCRO	UN Confidence Restoration Operation in Croatia
UNDOF	UN Disengagement Observer Force (Syria)
UNHCR	UN High Commission for Refugees
UNITA	*União Nacional para a Independência Total de Angola*
UNMIH	UN Mission In Haiti

UNOMIL	UN Observation Mission In Liberia
UNOSOM	UN Operation in Somalia
UNPA	UN Protected Area
UNPREDEP	UN Preventative Defence Force
UNPROFOR	UN Protection Force
UNSC	UN Security Council
UNSCOM	UN Special Commission on the Disarmament of Iraq
UNSCR	UN Security Council Resolution
UNTAES	UN Transitional Administration for Eastern Slavonia, Baranja and Western Sirmium
URNG	Revolutionary National Guatemalan Union
WEU	Western European Union
WHO	World Health Organisation
WTO	World Trade Organisation